ON THE DISTANCES BETWEEN SUN, MOON AND EARTH

POLISH ACADEMY OF SCIENCES
INSTITUTE FOR THE HISTORY OF SCIENCE,
EDUCATION AND TECHNOLOGY
CENTRE FOR COPERNICAN STUDIES

STUDIA COPERNICANA — BRILL'S SERIES

VOLUME I

EDITORS OF THE VOLUME:
JAN MALICKI AND JACEK SOSZYŃSKI

(POLISH ACADEMY SERIES XXX)

ON THE DISTANCES BETWEEN SUN, MOON AND EARTH

According to Ptolemy, Copernicus and Reinhold

BY

JANICE ADRIENNE HENDERSON

E.J. BRILL
LEIDEN • NEW YORK • KØBENHAVN • KÖLN
1991

The paper in this book meets the guidelines for permanence and durability of the Committee on Production Guidelines for Book Longevity of the Council on Library Resources.

Library of Congress Cataloging-in-Publication Data

Henderson, Janice Adrienne.
On the distances between sun, moon, and earth according to Ptolemy, Copernicus, and Reinhold / Janice Adrienne Henderson.
p. cm.—(Studia Copernicana; 30) (Studia Copernicana. Brill's series; v. 1)
Includes bibliographical references and index.
ISBN 90-04-09378-8 (alk. paper)
1. Cosmological distances. 2. Cosmological distances—Measurement—History. I. Title. II. Series. III. Series: Studia Copernicana. Brill's series; v. 1.
QB991.C66H46 1991
523.1—dc20 91-11833
CIP

ISSN 0925-6806
ISBN 90 04 09378 8

PRINTED IN THE NETHERLANDS

CONTENTS

CONTENTS (cont.)

CONTENTS (cont.)

LIST OF TABLES

LIST OF FIGURES

FOREWORD

With this volume, *Studia Copernicana* enter a new phase. On the basis of an agreement with E. J. Brill, Printers and Publishers in Leiden, we are starting a new series under the rubric *Studia Copernicana – Brill Series*, with volumes numbered in Arabic numerals, but with the current Roman numeration in brackets. Thus Dr Janice A. Henderson's book *On the Distances Between Sun, Moon and Earth according to Ptolemy, Copernicus and Reinhold* has the serial number 1 (XXX). It is published jointly by the Centre for Copernican Studies of the Institute for the History of Science, Education and Technology of the Polish Academy of Sciences in Warsaw, and Brill Publishers in Leiden, who are its sole distributors. The existing series will nevertheless continue as the 'Polish Academy Series', numbered as usual with Roman numerals. A list of titles published in both series will be found at the end of each volume. The Editor in Chief, in agreement with Brill Publishers, will decide in which series a given volume is to be included.

We are indebted to Dr Henderson for kindly allowing us to publish her dissertation of 1973 in *Studia Copernicana.* In a letter of 11 June 1988 she wrote: 'Please feel free to use [my thesis] in whatever way necessary for reproduction. I will not be making any changes to the text'. Her thesis, hitherto available only through University Microfilms International, but used and referred to by many scholars, will thus at last be accesible to all in book form.

Acknowledgements are due to the Ciechanowiecki Family Foundation of the Royal Castle in Warsaw for assuring financial assistance, which made this publication possible.

Professor John D. North of the International Academy of the History of Science has written the Preface to the book. It was he who initiated the project of cooperation between Brill Publishers and *Studia Copernicana*, approved and supported by Professor Jerzy Dobrzycki, Director of the Institute for the History of Science, Education and Technology. At E. J. Brill Publishers Mrs Elisabeth Erdman-Visser carried it into effect with enthusiasm and competence. Dr Jerzy A.Supel, of the Institute of Fundamental Technological Research in Warsaw, prepared the camera-ready pages for the printer, including the Figures, which were reproduced from the original typescript. The final proofs were corrected by Dr Grażyna Rosińska and Jarosław Włodarczyk. The editors of the volume, Jan Malicki and Jacek Soszyński, put all their creative skills into organising the computer-setting process and adjusting it to the publisher's requirements. All those mentioned here are to be thanked wholeheartedly for their devotion to the common task.

Paweł Czartoryski
October 1990

PREFACE

One might be excused for thinking that there are as many astronomers by the name of Copernicus as there are scholars writing on him, but whether they portray him as bold thinker or as timid canon, all must agree on this: that Copernicus has long been regarded as the quintessential intellectual revolutionary. One needs only to recall the eager response of studentes in the 1960s to lectures on the *De Revolutionibus* to appreciate the power of words, especially ambiguous words like 'revolution'. 'We revolve about the Sun like any other planet', Copernicus wrote in his *Commentariolus.* Whether or not he was the first to make the point, whether or not he was merely performing a simple geometrical transformation of Ptolemaic astronomy, these questions cannot detract from the simple truth that when he wrote such things almost no-one agreed with him, and that when – long after his death – a consensus was eventually reached, he was given almost universal credit for the change.

Historians of the exact sciences are often disdainful of what they are quick to represent as unmerited glory, or as simply a vulgar misconception, but it is surely in part precisely because Copernicus has this marked symbolic value for Western culture that it behoves us to look into the fine details of his astronomical writings. This Janice Henderson has done in many important respects. Taking as her starting point a manuscript by Erasmus Reinhold, she analyzes and compares the Ptolemaic and Copernican discussion of the sizes and distances of the sun and moon. She shows how, despite the fact that the Copernican lunar model was designed to correct a mistakenly exaggerated variation in Ptolemaic lunar distance, and so could hardly fail to be a step in the right direction, the observations he adduced in his own support were in all probability notional. Having reconstructed the essentials of Copernicus' work, she sees it as that of a master architect. For the correction of fine detail we must turn to the truly masterly apprentice, Reinhold.

This highly original study, completed seventeen years ago, has been long awaited. Its simultaneous publication in Warsaw and Leiden as the first of a new series (the Brill series of *Studia Copernicana*) may be taken to symbolize another revolution, one that has made such a venture possible.

J. D. North
October 1990

ACKNOWLEDGEMENTS

I would like to thank my initial thesis advisor, Professor Bernard R. Goldstein, who continued to assist me even after he had left Yale University. He has read the entire draft and has made innumerable suggestions. Professor Asger Aaboe gave unstintingly of his time in the final stages of preparation. His astute insight into technical problems, and his attempts to improve my prose style, have been invaluable to me. Professor Noel Swerdlow, in addition to originally suggesting the topic to me, has helped me with the translations of difficult passages and with illegible portions of Reinhold's manuscript. He has always been willing to puzzle over intractable numbers and in general has been a continual source of assistance and encourangement. Finally, I would like to thank my typist, Miss Christine M. Tattersall, who, as usual, has done a beautiful job with a difficult text.

I am indebted to the National Science Foundation for financial support while I was a graduate student.

Janice A. Henderson
1973

ABSTRACT

This work is a detailed analysis of one feature of the astronomy of Ptolemy and Copernicus – the sizes and distances of the sun and moon. I was led to this subject by my study of a manuscript of Erasmus Reinhold, discovered early in the present century, which has not previously been sufficiently well-investigated. It contains a brief commentary on the *Almagest*, dealing exclusively with the subject of the solar and lunar distances, and an extensive commentary on *De Revolutionibus*. While the latter commentary is primarily concerned with the derivation of the parameters which Reinhold used in constructing the *Prutenic Tables*, the section on the solar distance is an interesting exception.

The method for determining the solar distance used by Copernicus and Reinhold is modelled on the procedure described by Ptolemy in the *Almagest* and deriving ultimately from Hipparchus. It is a geometric method which depends on the apparent equality of the solar and lunar diameters, and requires previous knowledge of 1) the lunar distance at which solar and lunar diameters appear precisely equal, 2) the angular diameters of the sun and moon at that distance, and 3) the ratio of the radii of the moon and of the earth's shadow where it intersects the lunar orbit at that distance. Ptolemy's greatest and least lunar distances differ by a factor of almost two, which implies that the diameter and parallax of the moon would appear almost twice as large at the moon's closest approach to the earth as at its furthest distance. Copernicus' lunar model was specifically designed to correct this exaggerated variation in distance and parallax. Curiously, the parallax observations cited by Ptolemy and Copernicus serve to confirm their respective models. The apparent diameters of the sun, the moon, and the earth's shadow are obtained from observations of two lunar eclipses. While Copernicus follows the method of Ptolemy, I believe that his cited observations are fictitious.

I have reconstructed the details of Copernicus' work from his surviving holograph manuscript. His values for the apparent solar and lunar diameters and for the ratio of the lunar and shadow radii are merely adapted from the values of Ptolemy and al-Battānī. He begins his derivation of the solar distance with one set of parameters, neglecting to compute the lunar distance at which solar and lunar diameters appear equal, and arrives at a solar distance of 1179^{tr} . He later returned to this section of the manuscript, changed the parameters, and recomputed various intermediate values, but failed to adjust the final result. It appears from Reinhold's commentary on *De Revolutionibus* that he was dissatisfied with Copernicus' value for the solar distance. After three unsuccessful attempts to obtain a value, he assumes that the solar distance is 1200^{tr} and from this derives a value of the ratio of lunar and shadow radii. With this value he then computes a solar distance of 1208^{tr} , very close to Ptolemy's value in the *Almagest* of 1210^{tr} .

This thesis is largely devoted to a close analysis of the manuscripts of Copernicus and Reinhold. *De Revolutionibus* contains numerous computational errors and inconsistencies, many of them illuminated by the manuscript, and this had severe consequences for the usefulness of this work to astronomers. Reinhold's commentary, on the other hand, reveals that he was a meticulous computer who lavished painstaking care on even the smallest detail. Both Copernicus and Reinhold were much in awe of Ptolemy's authority and were unwilling to venture very far from his value of the solar distance.

INTRODUCTION

Soon after the publication of *De Revolutionibus*, certain astronomers who read Copernicus' work closely noted a series of errors.[1] They were in the form of inconsistencies, obvious computational errors, and discrepancies between positions computed from the tables in *De Revolutionibus* and observations. These astronomers used, of course, the edition of *De Revolutionibus* printed in Nuremberg in 1543, which I shall refer to as the Nuremberg edition. There is, however, a surviving holograph manuscript, discovered in Prague in the 1830's, which has twice been published in facsimile edition.This is not the manuscript from which the printed edition was prepared, but it serves to throw light on some of the problems in *De Revolutionibus.* In the manuscript, for example, one finds Copernicus frequently changing or correcting some, but not all, of the numerical values in a given computation, thus creating inconsistencies.

One of the careful readers of Copernicus' work was Erasmus Reinhold, the Wittenberg astronomer who compiled the *Prutenic Tables.*[2] Reinhold noted that the very observations in *De Revolutionibus* from which Copernicus derived his parameters and tables could not be reproduced from the tables themselves.[3] He undertook the task of rederiving all the parameters and recomputing the tables, an endeavor to which he devoted seven years.[4]

Even a cursory examination of the *Prutenic Tables* reveals the extent of Reinhold's revisions. He says in the preface that he brought together the observations of Copernicus with those of Ptolemy and others, and, aside from the bare observations and the traces ("vestigia") of the explanations, he took

---

[1] See, for example, the letter of Matthias Lauterwalt to Georg Rheticus, dated 17 February, 1545, in Burmeister [1],pp.59-67.

[2] Erasmus Reinhold was born on October 21, 1511, in Saalfeld in Thuringen, where he died of the plague on February 19, 1553. He spent most of his life in Wittenberg, where he was a mathematics professor at the University after 1536, a dean of the University in 1549, and rector in 1550. Reinhold's published works, in addition to the *Prutenic Tables* (Tübingen 1551), include his edition of and commentary on Peurbach's *Theoricae novae planetarum* (Wittenberg 1542), which was reprinted numerous times; an edition of the Greek text of book i of the *Almagest,* with a translation into Latin and a commentary by Reinhold (Wittenberg 1549); a short work on spherical astronomy, *Themata quae continent methodicam tractationem de Horizonte rationali ac sensibili...*(Wittenberg 1541); ephemerides for use in Wittenberg for the years 1550 -1551, *Ephemerides duorum annorum 50, et 51, supputatae ex novis tabulis astronomicis* (Tübingen 1550); and a trigonometric work, essentially a new edition of the *Tabulae directionum* of Regiomontanus, *Primus liber tabularum directionum...* (Tübingen 1554).

[3] Reinhold [1], α4r.

[4] Reinhold [1], α4r.

nothing else from Copernicus.[5] One of the original intents of my thesis was to extract Reinhold's parameters from his tables in order to verify this statement and to determine in exactly what sense the tables were Copernican. However, I soon learned that this would not be necessary, since Reinhold left a detailed account of his derivation of the parameters, preserved in a holograph manuscript entitled "Commentarius in opus revolutionum Copernici." [6]

There are three references to this comentary in the *Prutenic Tables.* It is mentioned in the privilege preceding the tables, dated 1549:

> Eruditus Commentarius in totum opus Revolutionum Nicolai Copernici.
>
> Learned commentary on the entire work of Revolutions of Nicolaus Copernicus.[7]

In the preface Reinhold says:

> Causas vero & rationem singularum compositionum exposui in Commentariis nostris, quos scripsi in opus revolutionum Copernici.
>
> I have explained the causes and the computation of the individual tables in our commentary which I wrote on the work of revolutions of Copernicus. [8]

There is a similar remark in the dedication:

> Itaque collatis Copernici observationibus cum Ptolemaeo & Hipparcho, alias tabulas institui quarum usum mox ostendam. Et causas earum ac fontes in alio opere expono.
>
> Comparing the observations of Copernicus with those of Ptolemy and Hipparchus, I constructed other tables, the use of which I will now set forth. And the causes and sources of these things I explain in another work. [9]

This work, evidently written before the completion of the *Prutenic Tables*, was thought lost already by the end of the sixteenth century. In 1596 Maestlin expressed regret that Reinhold was not able to produce the commentary which he had planned.[10] It was rediscovered in Berlin early in the present century by Valentin Rose and Ludwik A. Birkenmajer.[11] The complete manuscript[12]

---

[5] Reinhold [1], β2v.

[6] Reinhold [2].

[7] Reinhold [1], α2r.

[8] Reinhold [1], β2v.

[9] Reinhold [1], α4r.

[10] Kepler [1], p. 85.

[11] Birkenmajer [1], p. 172.

[12] The complete contents of the manuscripts, bound together in Berlin, Deutsche Staatsbibliothek, Latin 2° 391, are as follows: *Commentarius in opus Revolutionum Copernici* (books iii and iv) ff. 1-63; spherical trigonometry, June 4, 1543, ff. 64-124; theorems on

also contains an incomplete treatise on spherical trigonometry (dated June 4, 1543) and plane trigonometry, a commentary on a portion of the *Almagest* (dated January 1, 1549), and a commentary on book 2 of Pliny. Reinhold's name does not appear anywhere in the manuscript, but he is known to have lectured on the *Almagest* in 1549,[13] and agreement between the contents of the commentary and the *Prutenic Tables* leaves no possible doubt.

In the commentary on *De Revolutionibus* (Ms.R.28v),[14] Reinhold refers to a historical work of Jean Funck printed in Nuremberg in 1545,[15] which implies that the manuscript was written after that date. He also refers to "calculations of Ptolemy which I have written elsewhere for M. Johanne Aurifabra."[16] This occurs at the beginning of the section of the commentary on the lunar and solar distances, and corresponds exactly to the notes on the *Almagest* found in the Berlin manuscript. It seems unlikely, however, that the commentary could have been written as late as 1549, since the *Prutenic Tables* appeared in 1551.

The contents of the commentary on *De Revolutionibus* correspond to Reinholds's description in the *Prutenic Tables.* Taking the observations cited in *De Revolutionibus*, not only by Copernicus but by Ptolemy and Hipparchus as well, and using the Copernican models, Reinhold derives all the essential solar, lunar and planetary parameters. These parameters are precisely those used in the computation of the *Prutenic Tables.* There is some doubt as to the completeness of the surviving manuscript. It begins with a discussion of chapter 13 of book iii of *De Revolutionibus*, and covers the rest of book iii, as well as books iv and v. In general, it deals with the contents of Copernicus' work chapter by chapter, omitting those chapters which do not entail mathematical derivations. I belive this may well account for the absence of a commentary on the preliminary chapters of book iii. On the other hand, it seems quite likely that Reinhold would have commented on book iv, as this is material with which Reinhold is quite capable of dealing, and which is essential for the *Prutenic Tables.* There is also a section missing from the beginning of the derivation of

---

the theory of circles, ff. 125-171; On Ptolemy's *Almagest* book v (chapters 12-16), January 1, 1549, ff. 172-186; the remainder of the commentary on *De Revolutionibus* (book v), ff. 192-294; *Tabulae in capita mathematica Plinii*, ff. 295-324; an astrological work in a different hand, dated 1575, ff. 325-356. The manuscript is described in Rose [1], pp. 1366-1367. At least one other manuscript of Reinhold's survives, also in Berlin, Deutsche Staatsbibliothek, Latin 4° 32. This contains lecture notes on Euclid, dated between November 1543 and August 1550.

13 Zinner [1], p. 513.

14 For this and similar abbreviations, see the definition of terms following the introduction.

15 Birkenmajer [1], p. 173.

16 Ms.R.50v. There are two contemporaries of Reinhold named Joannes Aurifaber, both of whom studied at Wittenberg. The first Joannes (1519-1575), from Mansfeldt, was Luther's private secretary known for his publication of Luther's *Letters*. The second Joannes (1517-1568), from Breslau, is probably the one to whom Reinhold refers. He graduated from the University of Wittenberg in 1538, was a docent there until 1550, and was a lifelong friend of Melanchton, Reinhold's colleague and supporter. Cf. Anon. [1].

Jupiter's parameters, where Reinhold has indicated in the manuscript "Initium vide in alio libro" (Ms.R.208r).

Reinhold's commentary on *De Revolutionibus* contains nothing original of a purely theoretical sort, nor does it provide insight into the obvious technical problems of Copernicus' work. Using the models, observations and mathematical methods of *De Revolutionibus*, Reinhold displays in detail the computation of each parameter. His results differ from those of Copernicus because he is a careful, meticulous computer. He uses seven and sometimes eight place trigonometric functions, and he carried all the digits throughout each computation. He very rarely makes an arithmetical mistake. Copernicus, on the other hand, made numerous computational errors which affect the parameters in *De Revolutionibus*.

One section of Reinhold's commentary on *De Revolutionibus* which I found of particular interest was his analysis of the sizes and distances of the sun and moon. Reinhold's normal procedure is to begin with the observations stated by Copernicus and derive the parameters as Copernicus should have done it. His treatment of the solar distance, however, is the only portion of his commentary which is not strictly modelled on *De Revolutionibus*. It appears that Reinhold was not satisfied with the solar distance resulting from Copernicus' data, and he altered the data in order to arrive at a solar distance close to Ptolemy's value.

Reinhold's copy of the *Almagest* was undoubtedly the Basel 1538 edition of the Greek text.[17] He published an edition of book i of the *Almagest*, identical to the Basel edition, with a translation into Latin and a commentary, in 1549. Reinhold's commentary on the *Almagest* covers only book v, chapters 12 to 16, which deal precisely with the sizes and distances of the sun and moon. It appears to be a selfcontained unit, with an epilogue, and book v may be the only portion of the *Almagest* on which he commented.

Reinhold, of course, was using only the Nuremberg edition of *De Revolutionibus*. The section of Copernicus' manuscript on the solar distance proves to be very interesting in itself. Copernicus has used two different sets of parameters, adapting the values of Ptolemy and al-Battānī, and conflated the results, making frequent errors along the way. It turned out to be possible to reconstruct most of his assumptions and computations, though not without some difficulty. My analysis of Copernicus' determination of the solar distance, which depends on the lunar distance and is modelled on Ptolemy's method, and of Reinhold's commentary on the determination of the lunar and solar distances in both the *Almagest* and *De Revolutionibus*, became, then, the major subject of my dissertation.

The solar distance played a significant role in the cosmology of both Ptolemy and Copernicus. The Ptolemaic planetary models, described in Ptolemy's *Almagest*, implicitly contain the heliocentric distances of the planets relative to the distance of the sun from the earth. For an inner planet, if the radius of

[17] Ptolemy [2].

the deferent is R, the radius of the epicycle gives the distance of the planet from the sun in terms of R. For an outer planet, the heliocentric distance is measured by the reciprocal of the epicyclic radius. Ptolemy accounted for the planetary distances, however, by the theory of nested spheres, described in his *Planetary Hypotheses*.[18] This theory assumes that the least distance of a planet is equal to the greatest distance of the planet just below it, and that the ratio of the relative distances of a planet, resulting from the model, is equal to the ratio of the true distances. Ptolemy accepts the order of the planets given by the ancients (*Almagest* ix.1) – Moon, Mercury, Venus, Sun, Mars, Jupiter, Saturn, Fixed Stars. There was disagreement among astronomers, however, about whether Mercury and Venus were above or below the sun. From the lunar distance in terrestrial radii, Ptolemy obtains the true solar distance by a geometric method which is described in detail in this thesis. Using the ratios of the greatest to least distances of Venus and Mercury produced by their models, he determines that these two planets fit nicely between the moon and the sun, leaving but a very small space in which Mars certainly does not fit. Ptolemy sees the solar distance as confirming his theory of nested spheres.

In the Copernican system, the true distances of the planets in terms of the geocentric distance of the sun are established for the first time. We realize that the solar distance gives us the absolute scale of the solar system, but Copernicus appears not to draw the obvious conclusion. While recognizing the advantage of his system (*De Revolutionibus* i.10), to my knowledge he does not utilize this feature of heliocentricity to compute the true distances of the planets.

The solar distance, or the astronomical unit as it is also called, continues to be an important parameter which is used for measuring the dimensions of the solar system and of the universe. In the eighteenth century, elaborate expeditions were organized to observe the transits of Venus in 1761 and 1769, which were to be used for the express purpose of determining the solar distance. In the seventeenth century, a reasonable approximation of the solar distance had been obtained by estimations of solar parallax, though these estimates differed by as much as a factor of two. Previous to this time, Ptolemy's estimate of the solar parallax, which was in error by a factor of about 20, had been accepted as essentially correct. Copernicus and Reinhold, in particular, were not prepared to change Ptolemy's value for the solar distance substantially, as we shall see.

More specifically, my thesis is an analysis of the geocentric distances of the sun and moon as found in the *Almagest*, in both the manuscript version and the Nuremberg edition of *De Revolutionibus*, in Reinhold's commentary on the *Almagest*, and in his commentary on this section of *De Revolutionibus*. Chapter one contains a detailed analysis of the lunar distance, and chapter two concerns the apparent diameters of the sun, moon and shadow. The Ptolemaic method, which is the model for Copernicus and Reinhold, requires the determination

[18] Goldstein [1].

of these quantities as a preliminary to the calculation of the solar distance, which is treated in chapter three. The fourth chapter is a brief analysis of the relative magnitudes of the sun, moon and earth, which Ptolemy, Copernicus and Reinhold discuss after they have reached values of the lunar and solar distances. The final chapter concerns an application of the distances – the solar and lunar parallaxes and diameters. A more detailed summary of the results obtained by Copernicus and Reinhold is given below, following a brief history of the determination of the solar distance.

The earliest surviving attempt to determine the solar distance is that of Aristarchus of Samos.[19] His method was to measure the angle subtended at the earth by the moon and sun when the moon appeared exactly halved. In the triangle formed by sun, moon and earth, with a right angle at the moon, the cosine of the angle at the earth is the ratio of the lunar distance to the solar distance. Both the measurement of this angle, which is very close to ninety degrees, and the determination of the moment when the moon is exactly halved, are very difficult to perform. The result found by Aristarchus, that the solar distance is between 18 and 20 times the lunar distance, is cited by Archimedes in the *Sand-Reckoner*.[20] Archimedes himself estimated the solar distance as 30 times the lunar distance, although his estimate appears not to be based on any geometric method. It is difficult to tell how seriously Archimedes should be taken, since he is interested in over-estimating the size of the universe.

Hipparchus invented a new geometric method relating the distances of the sun and moon such that when one is given the other can be calculated. Given the ratio of the solar and lunar distances, there remains the problem of determining one of them in absolute terms, i.e. in earth radii. This can be done by a direct observation of parallax, since the reciprocal of the sine of the parallax of an object seen on the horizon is equal to the distance of the object in terrestrial radii. Parallax is measured by comparing the apparent position of an object with the true position predicted by theory, and thus requires a theory wich can adequately predict true positions. The lunar distance can be obtained with some degree of accuracy from the parallax of the moon, but solar parallax is so small that the distance of the sun cannot be determined in this way by naked-eye observation. Hipparchus, however, appears not to have attempted a direct measurment of either lunar or solar parallax. Ptolemy and Pappus report that Hipparchus *assumed* values for the solar distance and from these derived two different sets of lunar distances, presumably arising from the two different assumptions about the solar parallax.[21] This may indicate that Hipparchus did not have sufficient confidence in the accuracy of his lunar theory to use it for a direct measurement of parallax. Indeed, his lunar model may

[19] Heath [1].

[20] Archimedes [1], p. 223.

[21] Swerdlow [2], pp. 287-290.

have been adequate for syzygy, and thus for the prediction of eclipses, but not for positions of the moon at an arbitrary elongation from the sun. Hipparchus' value for the solar distance was 490^{tr}, not 2490^{tr} as Hultsch has argued.[22]

Ptolemy used the geometric method of Hipparchus, but derived the solar distance from the lunar distance. For Ptolemy, and for those who followed his example, which includes al-Battānī, Copernicus and Reinhold, the lunar distance and the lunar model were essential features of the determination of solar distance. Ptolemy made several changed in the lunar model of Hipparchus to improve its accuracy at positions other than syzygy. In particular, he had noted that the simple model of Hipparchus produced longitudes near quadrature which did not agree with observations. Ptolemy's complex lunar model accounts for these deviations by bringing the center of the lunar epicycle closer to the earth at quadrature, increasing the apparent size of the epicycle and thus the effect of lunar anomaly. The ratio of Ptolemy's maximum lunar distance at syzygy to the minimum distance at quadrature implies that the diameter and parallax of the moon would appear almost twice as large at the moon's closest approach to the earth as at its furthest distance. While nothing of the kind should occur, and Ptolemy makes no reference to this problem, the parallax measurement reported in the *Almagest* curiously confirms the close approach of the moon to the earth near quadrature.

Copernicus' lunar model was specifically designed to correct the exaggerated variation in distance and parallax produced by Ptolemy's model without impairing the quality of its predictions of lunar longitudes at quadrature, and was in fact successful in this respect. His double epicyclic model increases the equation of anomaly at quadrature while keeping the center of the epicycle at a constant distance, and thus the relative distances remain within reasonable limits. Copernicus describes two lunar parallax observations in *De Revolutionibus* iv.16, both made near quadrature. Though the computations are filled with errors, the results effectively disprove the close approach of the moon to the earth implied by the Ptolemaic model.

Between the time of Ptolemy and Copernicus, the only attempt well-known in the West to redetermine the solar distance was that of al-Battānī.[23] The geometric method of Hipparchus and Ptolemy requires the previous knowledge of

1. the lunar distance at which the solar and lunar diameters appear equal,
2. the angular diameters of the sun and moon at that distance,
3. the ratio of the radii of the moon and the shadow of the earth where it intersects the lunar orbit at that distance.

Ptolemy found that the solar and lunar diameters appeared equal when the moon was at its greatest distance at syzygy. He determined this distance in

[22] Swerdlow [2], p. 290.

[23] Al-Battānī (c. 850-929) was born in Batan in Mesopotamia and died in Damascus. For al-Battānī's derivation of the solar distance, see Nallino [1], v.1, pp. 57-61.

terrestrial radii, found the angular diameter of the moon at this distance, which was then equal to the angular diameter of the sun, and found the angular diameter of the earth's shadow at this distance. Al-Battānī adopted Ptolemy's lunar model, lunar distance, solar diameter, and the ratio of lunar and shadow radii. He redetermined the apparent diameter of the moon, however, which necessitated recomputing the lunar distance at which solar and lunar diameters appear equal. If al-Battānī had used his value for this lunar distance to find the solar distance, he would have found that his parameters produced a solar distance 200 times greater than that of Ptolemy, a result which may well have seemed unreasonable and surprising. He circumvents this problem by simply multiplying his lunar distance by the ratio of Ptolemy's solar and lunar distances, obtaining a result closer to Ptolemy's namely 1146^{tr}.

Copernicus appears to have original values for the apparent solar and lunar diameters and for the ratio of the lunar and shadow radii. I believe, however, on the basis of my analysis, that he has merely adjusted the values of Ptolemy and al-Battānī to fit the lunar distances which arise from his new lunar model. Copernicus' solar diameter at apogee follows from Ptolemy's value of the solar diameter, and his lunar diameter follows from al-Battānī's value of the lunar diameter. There are two different values for the ratio in Copernicus' manuscript. The first is an adjustment of Ptolemy's value of the ratio. Copernicus, it must be understood, does not reveal the source of his parameters, but merely states them as if he had determined them anew. In fact, he has done nothing of the kind. This is particularly apparent in his derivation of the radii of the moon and the shadow. I believe that the two observations which Copernicus says he used (*De Revolutionibus* iv.18) were fictitious.

From the manuscript of *De Revolutionibus* one can see that Copernicus began his derivation of the solar distance with a value of 79:30 for the ratio. He did not find the lunar distance at which solar and lunar diameters appear equal, but merely used the greatest distance at syzygy to derive a solar distance of 1179^{tr}. Later, he returned to this section of the manuscript and changed the value of the ratio to 403:150. At this time, he also computed the lunar distance at which solar and lunar diameters appear equal. While he recomputed various intermediate values in the derivation, he did not bother to adjust the final result. The Nuremberg edition of *De Revolutionibus* contains the value 403:150 and the correctly computed lunar distance, but retains the solar distance of 1179^{tr}, which thus appears not to follow from the stated parameters.

Both al-Battānī and Copernicus relied heavily on Ptolemy, and I would venture to say that they wished to obtain a solar distance which differed but little from that of Ptolemy. This is even more explicit in the work of Reinhold. The solar distance itself is not constant, but varies between a greatest distance in apogee and a least distance in perigee. In the *Almagest*, Ptolemy considered only one solar distance, 1210^{tr}; since he nowhere in the *Almagest* specified that this was the mean distance, as he does in the *Planetary Hypothesis*, later astronomers assumed it to be the greatest distance. Al-Battānī and Copernicus, by estimating the solar diameter at apogee and determining the lunar distance at which solar and lunar diameters appear equal, found the solar distance in apogee. The Copernican solar model, with its slowly changing eccentricity, introduces a further variation in distance such that there is actually a greatest and least solar distance in apogee and perigee. Copernicus himself does not consider the effects of the changing eccentricity on the solar distance, but Reinhold treats it at length.

According to Copernican theory, the sun was at its greatest distance in apogee near the time of Hipparchus and Ptolemy. In order to find the greatest solar distance in apogee at this time, Reinhold uses Ptolemy's value for the solar diameter. He accepts Copernicus' new value for the lunar diameter, presumably an unchanging quantity redetermined correctly by Copernicus. With these two diameters, and using his recomputed elements of the Copernican lunar model, Reinhold finds the lunar distance at which the lunar diameter appears equal to the solar diameter in apogee. His values for the greatest and least lunar distances at syzygy and quadrature differ slightly from those of Copernicus, both because Reinhold has redetermined the parameters of the lunar model, and thus found different relative distances, and because he has more carefully reduced the lunar parallax observation used to convert the relative lunar distances into absolute distances in terrestrial radii.

For the final variable in the determination of the solar distance, the ratio of the lunar and shadow radii, Reinhold begins by assuming Ptolemy's value for the ratio (13:5) and obtains a solar distance nearly three times as large as that of Ptolemy (3108^{tr}). He remarks that this value is too large and that Copernicus was correct in assuming a different value for the ratio. Using Copernicus' value of 403:150, Reinhold finds a solar distance of 1301^{tr}, which he considers still too large. He makes one more attempt, with a ratio of 19:7, obtaining 1187^{tr}. While this value is close to that of Copernicus, he is obviously still not satisfied with the result. Reinhold's next step is to assume that the solar distance is 1200^{tr}, a value very close to Ptolemy's 1210^{tr}, and from this he derives a value for the ratio, 122:45.

Finally, with this value of the ratio, he computes a solar distance of 1208^{tr}, which is even closer to Ptolemy's value. Theoretically, he should have obtained the solar distance he originally assumed, but due to a computational error he does not. This seems to me a telling example of the strong influence of Ptolemy, even after the time of Copernicus.

Ptolemy's influence on Copernicus is much in evidence throughout *De Revolutionibus*, which is modelled very closely on the *Almagest*. Not only does Copernicus adhere to the subject matter of the *Almagest*, presented in roughly the same order, but his mathematical methods, used in the derivation of parameters and the construction of tables, are virtually identical to Ptolemy's. The determination of the solar distance is but one example of Copernicus' use of the actual numerical values in the *Almagest*. The underlying concern throughout *De Revolutionibus* is to show that the Copernican models produce results which agree with the Ptolemaic models.[24]

One of the reasons the Ptolemaic system survived without rival until the time of Copernicus is the high quality of Ptolemy's workmanship. The *Almagest* is a model of logical rigor, both in its presentation of the subject matter and in the accuracy of the numerical details. As for the observations, John Britton has suggested, on the basis of his analysis of solar and lunar observations and parameters in the *Almagest*, that Ptolemy's procedure for deriving the parameters from the observations may be considerably more sophisticated than is indicated in the *Almagest.* Since Ptolemy's parameters are a great deal more accurate than one would expect on the basis of his method for deriving the parameters and on the basis of the average errors in his observations, it appears that he may have been drawing on a large number of observations.[25] After many determinations of the parameters, and perhaps applying some non-rigorous averaging procedure, he may have selected the values which he thought best. There is nothing comparable to this in the work of Copernicus and Reinhold, who limit themselves to the procedures explicitly described in the *Almagest.*

In treating one small aspect of Copernicus' work – the solar and lunar distances – I am interested in the quality of his work as it would be judged by serious astronomers of his time. I am not concerned with the relation between the solar distance and the dimensions of the universe, nor with the philosophical implications of heliocentricity. Apart from the non-technical discussions in book i, the issue of heliocentricity is of no particular significance in *De Revolutionibus.* It is certainly irrelevant to the construction of tables, which, if they are to be at all practical, must be designed for an observer on the earth. One cannot determine from the *Prutenic Tables*, for example, whether Reinhold subscribed to a heliocentric system, and Reinhold gives no indication, either in his printed works or in his manuscripts, of his opinion of the heliocentric theory of Copernicus. There are two passages in his commentary on *De Revolutionibus* which seem to imply Reinhold's adherence to a Tychonic system, but in my opinion, the evidence is slim and hardly sufficient for a definitive assessment of his philosophical outlook.

What I have learned from my analysis of Copernicus' and Reinhold's work is the marked contrast between the two men in their treatment of numerical detail. Copernicus is quite unreliable and uneven in this respect. In the chapter on the solar and lunar distances, one has many examples of Copernicus altering his observational data, committing computational errors, and making partial corrections which are not carried through, resulting in internal inconsistencies in the printed editions of *De Revolutionibus.* I believe Reinhold was well aware of Copernicus' lack of concern for the arithmetical precision in his work, and that this accounts for Reinhold's decision to recompute all parameters before constructing the extensive *Prutenic Tables*, since nothing short of a complete

---

[24] Neugebauer [2], p. 103.

[25] Britton [1], p. 204.

recomputation could rectify the situation. Reinhold's commentary, on the other hand, reveals the painstaking care which he lavished on even the smallest detail. I have included in the following some transcriptions of Reinhold's manuscript, along with translations, but only from the sections on the solar and lunar distances. The entire manuscript should be extensively analysed, particularly in connection with the *Prutenic Tables*, and this will be forthcoming.

As for the solar distances, my analysis of Copernicus' manuscript reveals that he was particularly careless in the computation of this basic dimension of the solar system. He did not independently derive the parameters which he used, but merely modified the values of Ptolemy. This has the advantage, of course, of producing a solar distance which does not differ greatly from that of Ptolemy, which was undoubtedly Copernicus' intent. There is no doubt about Reinhold's desire to derive a solar distance approximately equal to Ptolemy's. When he assumes that the solar distance is 1200^{tr}, he explicitly says that this does not differ too much from Ptolemy's value. To my knowledge, this is the only example in the commentary on *De Revolutionibus* where Reinhold rejects Copernicus' data. Since he was then forced to rely on Ptolemy, perhaps this is why he wrote a commentary on this very subject as it is treated in the *Almagest*. Their unwillingness to venture very far from the solar distance of Ptolemy is a good example of the extent to which both Copernicus and Reinhold were still in awe of his authority.

Definition of Terms

In general, by the distance of an object from the earth I mean the distance between the center of the earth and the object. When the distance is measured from an observer on the surface of the earth, I clearly so specify. The "relative" distance of an object, such as the moon, is measured in terms of a unit which is an element of the model. Ptolemy, for example, designated the mean lunar distance in syzygy as 60 units and his relative distances for other positions of the moon are then normed to this unit. Copernicus' relative lunar distances are given in relation to a deferent radius of 100 000 parts. The unit of measurement for a"true" distance, on the other hand, is one of fixed and unvarying length. Ptolemy and Copernicus obtain the distance of the moon, and from this the distance of the sun, by solving a triangle one side of which is the radius of the earth. Throughout this thesis, true distance refers to a distance measured in terrestrial radii, abbreviated "tr."

A distinction is made between "apparent" and "corrected" time. Apparent time, generally given in connection with an observation, is measured relative to local apparent noon, i.e., the moment of solar culmination (true noon). Corrected time is measured relative to mean noon and differs from the apparent time by the equation of time. For Ptolemy, there is an additional distinction between "equinoctial" and "seasonal" hours. The length of daylight, which varies with the time of year, is divided into 12 equal seasonal hours. Similarly, the time from sunset to sunrise is divided into 12 seasonal hours. Thus for a given day, unless it is the day of an equinox, the length of a seasonal hour before or after midnight is not equal to a seasonal hour before or after noon. In one complete solar day of unvarying length there are 24 equinoctial hours, and all equinoctial hours are equal.

The celestial sphere provides a convenient and natural reference system for the directions to celestial objects from an observer on the earth. The celestial sphere is a spherical surface of unit radius, arbitrarily large, with center at the observer's eye. The point of intersection of the celestial sphere with a line of sight to an object defines the direction to the object. By the position of the object we usually mean the position of its image on the celestial sphere.

If the center of the earth is taken as a standard viewing point, the geocentric (or daily) parallax is defined as the angle between the lines connecting the object with the center of the earth and an observer. The daily parallax varies as a function both of the observer's latitude and of time of day.

The term "true", when applied to celestial positions, is unfortunately used in two different senses. An"apparent" position of a celestial object refers to the direction in which the object is seen by an observer on the surface of the earth. A "true" position refers to a position derived from the model, which assumes that the object is seen from the center of the earth. The difference between apparent and true positions should thus be the daily parallax. Strictly speaking, this is correct only if the position predicted by the model does in fact coincide with the direction in which the object would be seen from the center of the earth. The fixed stars show no annual parallax, let alone daily parallax, so that they, or rather their images on the celestial sphere, may be used as a coordinate system for comparing true and apparent positions. The terms "apparent" and "observed" are used interchangeably. The term true as opposed to "mean" longitude is defined in chapter one in connection with solar and lunar longitudes. Here the term "true" does not distinguish between computed and observed positions, since both true and mean longitudes are derived from the model. The use of the word "true" is highly misleading and ideally should be avoided altogether. Unfortunately it is the common terminology of both Ptolemy and Copernicus.

Throughout the thesis, references to the *Almagest* and to *De Revolutionibus* are given by book and chapter, e.g., *Almagest* v.15 refers to book five, chapter fifteen. A similar system is used for works such as Ptolemy's *Geography*, and references to Euclid's *Elements* and Regiomontanus' *Epitome* are given by book and proposition number. References to Reinhold's commentary on *De Revolutionibus* and the *Almagest* are given by the folio number of the manuscript, e.g., Ms.R. 139v indicates folio 139 *verso*.

The determination of solar and lunar distances involves numerous calculations, both of solar and lunar positions and of geometrically determined distances and sizes. Reinhold's commentary on Ptolemy and Copernicus contains recomputations of all the numerical values. I myself have redone all the computations discussed in this thesis to check the calculations of Ptolemy, Copernicus and Reinhold. The complete sets of numerical values for any computation, i.e., the initial values, the intermediate steps and the final results, are listed in tables throughout the thesis under the headings Ptolemy, Copernicus and Reinhold. In general, only the final results and discrepancies between my recomputations and the values given by the three astronomers are discussed in the individual chapters. The tables are arranged so that Reinhold's recomputations can be compared with the original results of Ptolemy and Copernicus. Values which I have reconstructed and which do not appear in the *Almagest*, *De Revolutionibus*, or Reinhold's manuscript are bracketed in the tables. In the tables containing Copernicus' values, I have attempted to indicate changes which he made in the manuscript. In those columns which contain the values found in the Nuremberg edition of *De Revolutionibus*, an arrow before a number indicates that the value differs from that found in the Copernican manuscript.

CHAPTER ONE

THE DISTANCE OF THE MOON

The Ptolemaic and Copernican determinations of the solar distance require the preliminary derivation of the distance of the moon. In this chapter, I shall be concerned with the lunar distances found by Ptolemy and Copernicus, and with Reinhold's recomputation of their results. Each step in the procedure is first described in general and then the specific results are analysed in detail. In particular, all numerical values have been recomputed and discrepancies between the results given in the texts of Ptolemy, Copernicus and Reinhold and my recomputed values are pointed out. In the case of Ptolemy and Reinhold, there are very few errors. In analysing Copernicus' work, however, there are two types of problems. One concerns the simple arithmetical errors, which are far more frequent in the computations of Copernicus than in those of Ptolemy and Reinhold. Secondly, problems arise in analysing the original manuscript of *De Revolutionibus.* Here one finds Copernicus frequently changing his numerical values, either by altering the way in which they are written or by striking them out and replacing them with new values. An attempt has been made to account for these changes wherever possible.

The procedure for finding the lunar distances is as follows:

1. one observes the apparent zenith distance of the culminating moon under certain specified conditions,
2. one computes the true zenith distance of the moon at the time of the observation,
3. the parallax is then the difference between the true and apparent zenith distances,
4. the true lunar distance in terrestrial radii at the time of the observation is determined from the parallax,
5. the relative lunar distance at the time of the observation is found from the model,
6. the ratio of true and relative distances is then used to convert the extreme distances of the model, i.e., greatest and least distances in syzygy and quadrature, into true distances in terrestrial radii.

The observation itself merely involves a statement of the time of the observation and the apparent zenith distance. Ptolemy cites an observation he himself made in A.D. 135, when the moon was within 12° of quadrature. Copernicus describes two observations of lunar zenith distance, one just before and one just after the moon was at first quarter. In his report of the first observation, the time has twice been altered in the manuscript. For the second observation, three different values of the observed zenith distance are given in the manuscript, plus a fourth value in the Nuremberg edition. It does not appear that these changes were made in an effort to produce preconceived results. Some changes are too small to appreciably affect the results, while others have an adverse effect on the lunar distance, assuming Copernicus' aim was to maintain Ptolemy's lunar distance in syzygy while increasing the distance in quadrature.

The computation of the true zenith distance is rather lengthy and involves many intermediate steps. Reinhold gives almost all of the intermediate values in his manuscript, but for Ptolemy and Copernicus it has been necessary to reconstruct many of the intermediate steps. Detailed analyses of all the numerical values in the zenith distance observations of Ptolemy and Copernicus, plus Reinhold's recomputation, are given in Appendix A. The complete set of numerical values are listed in the tables in chapter one and only discrepancies and final results are discussed in detail.

The computation of lunar parallax from the true and observed zenith distances involves only a simple subtraction. I would venture to say that it is at this point Copernicus makes the changes in his apparent zenith distances, i.e., after obtaining the true zenith distance, he altered the apparent zenith distances to produce different values of the parallax.

The true lunar distance at the time of the observation is a function of the parallax and the apparent zenith distance. For Copernicus and Reinhold, there is a direct solution, applying the law of sines to the triangle through the moon, the observer, and the center of the earth. The three known elements in the triangle are the parallax, the apparent zenith distance, and the terrestrial radius, assumed to be one. Ptolemy's solution is somewhat longer, since he divides the triangle into two right triangles and solves each triangle separately.

The determination of the relative lunar distance requires a lengthier geometric procedure. Here the distance is a function of the parameters of the lunar model and the particular configuration of the model at the time of the observation. While Copernicus found the true lunar distance for both his observations, he finds the relative lunar distance for his second observation only.

The relative lunar distance at any given moment can be determined from the model. Given the values of the true and relative distances for any one configuration, the ratio of these distances can be used to convert relative distances into true distances. The greatest, least and mean distances in syzygy and quadrature are listed in tables at the end of chapter one. In the section of the *Almagest* on lunar distance, Ptolemy explicitly mentions only the mean distances in terrestrial radii. However, the greatest and least distances are obtained from the mean distances by a simple addition or subtraction of the epicycle radius, which Ptolemy does convert into terrestrial radii. In the Copernican lunar model, the mean distances in syzygy and quadrature are equal. Copernicus converts and mentions all extrema distances of the moon.

His mean distance, $60;18^{tr}$, is close to Ptolemy's value of the mean distance in syzygy, 59^{tr}. Ptolemy's greatest and least lunar distances, $64;10^{tr}$ and $33;33^{tr}$, differ by a factor of almost two, while Copernicus' greatest and least distances are within $8;2^{tr}$ of his mean distance.

1. PARALLAX AND ITS MEASUREMENT

The daily or geocentric parallax of an object is the difference between the true position of the object, as it would be seen from the center of the earth, and the apparent position of the object, as seen by an observer on the surface of the earth. In Figure 1.1, C is the center of the earth, O is an observer on the surface of the earth, and ZAB is the celestial sphere, with center O. The apparent position of an object at M is the point B on the celestial sphere. The true position of the object lies in the direction CM. OA, parallel to CM, is the direction in which the true position is seen by the observer. The daily parallax of the object at M is the arc AB on the celestial sphere, equal to $\angle AOB$. In practice, this means that while the true position of the object lies in the direction OA, to the observer the object will appear to lie in the direction OB. The effect of parallax is to increase the distance of the object from the observer's zenith Z.

For the observer, the most convenient reference system for determining parallax is the local coordinate system of altitude and azimuth. If the object is at the observer's zenith, its parallax is zero. Assuming a fixed distance from the earth, parallax in altitude increases with distance from the zenith, reaching a maximum when the object is on the horizon. When $\angle MOC$ is 90°, the parallax is called horizontal parallax. Further, parallax does not affect an object's azimuth.

In Figure 1.1, since OA is parallel to CM, $\angle AOB$ is equal to $\angle OMC$. Thus parallax is also the angle subtended at the object by the radius of the earth terminating at the observer. The closer the object is to the earth, the larger this radius will appear or, more precisely, the parallax (or rather its sine) is inversely proportional to the distance.

If π is the horizontal parallax, the distance is simply

$$S = \frac{1}{\sin \pi},$$

if the terrestrial radius is 1.

The moon alone of all heavenly bodies is so close to the earth that its daily parallax is directly observable with the naked eye, and thus it is possible to determine its distance. In Figure 1.2, O is the observer, C is the center of the earth, COD is the direction of the zenith, and M is the moon. The apparent zenith distance of the moon, z', is $\angle DOM$, and the true zenith distance, z, is $\angle DCM$. The parallax of the moon, π, is OMC, and

$$\pi = z' - z, \tag{1.1}$$

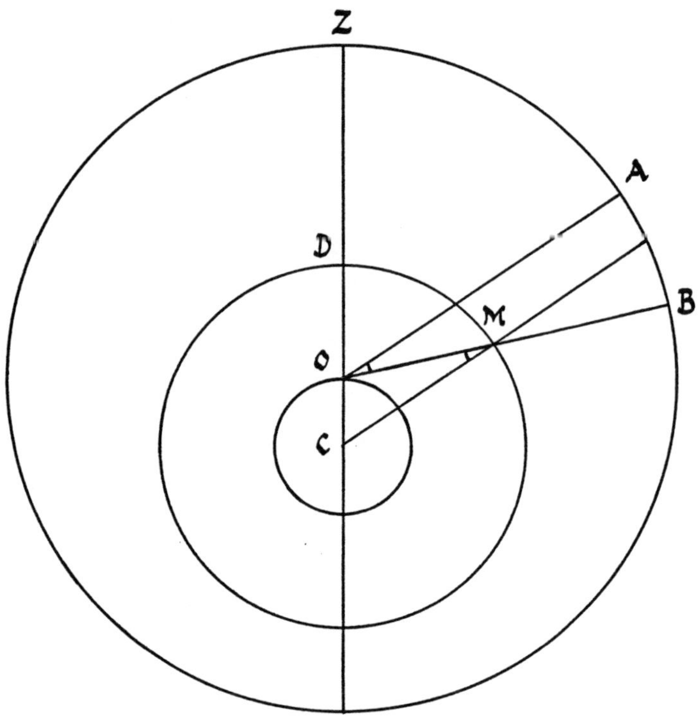

Figure 1.1: Daily Parallax

i.e., parallax is the difference between apparent and true zenith distance.

The apparent lunar zenith distance may be obtained by direct observation. The true zenith distance is found by computing the lunar longitude and latitude from the model at the time of the observation, and transforming these coordinates into altitude and azimuth. This procedure can be simplified if the culminating moon is observed when the pole of the ecliptic is in the meridian. Then the zenith distance is

$$z = \phi \pm \delta_m^* \pm \beta \tag{1.2}$$
$$= \phi \pm \delta_m, \tag{1.3}$$

where ϕ is the terrestrial latitude of the observer, δ_m^* is the declination of the point on the ecliptic which has the same longitude as the moon, δ_m is the declination of the moon, and β is the latitude of the moon. The signs of δ_m^*, δ_m and β depend on the position of the moon relative to the equator and the ecliptic.

Once the parallax has been determined, the distance of the moon can be found as a function of the parallax and either the true or apparent zenith distance. If S is the geocentric lunar distance, and the terrestrial radius is

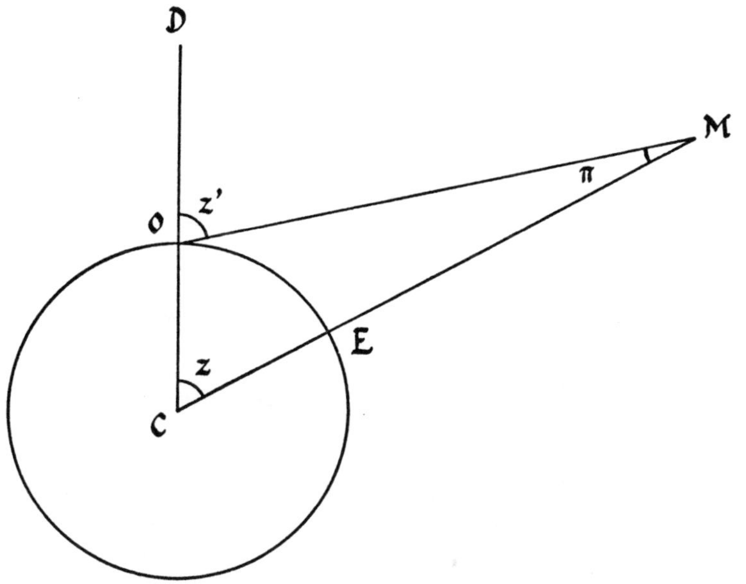

Figure 1.2: True and Apparent Lunar Zenith Distances

taken as unity, then

$$S = \frac{\sin z'}{\sin \pi}.$$

This is the relationship used by Copernicus and Reinhold to determine the lunar distance. In modern notation, Ptolemy's formula for the lunar distance is essentially

$$S = \sin(90^o - z) + \sin(90^o - \pi)\frac{\sin z}{\sin \pi}.$$

This is equivalent to

$$S = \cos z + \frac{\sin z}{\tan \pi}$$

or

$$\tan \pi = \frac{sinz}{S - \cos z}.$$

*1.1 The Parallactic Instrument (*Almagest *v.12,* De Revolutionibus *iv.15)*

Both Ptolemy and Copernicus describe in some detail the construction of an instrument for observing zenith distances. Ptolemy implies that he built his own parallactic ruler,[1] but his wording leaves some doubt as to whether he was actually the inventor.[2] The parallactic instrument used and allegedly built

[1] The parallactic ruler or instrument was known to the Arabs as an "alhidada longa," and in the middle ages as a "triquetus." Cf. Rome [1], pp. 69-77.

[2] Manitius [1], p. 33.

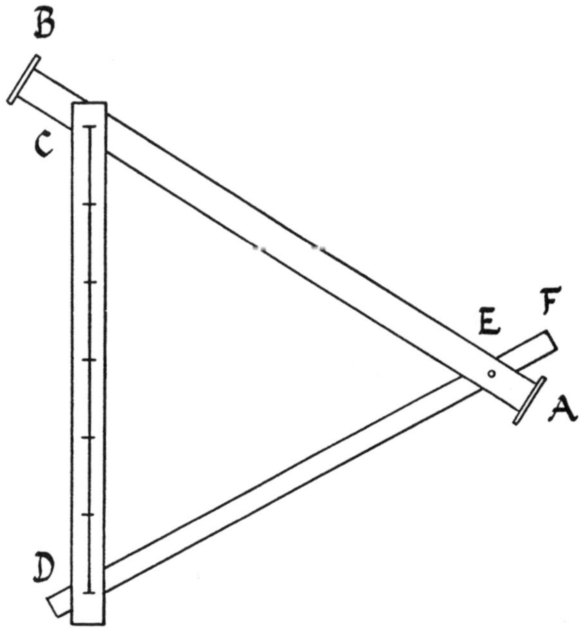

Figure 1.3: Ptolemy's Parallactic Instrument

by Copernicus was in the possession of Tycho Brahe, who describes it in his *Astronomiae Instauratae Mechanica*.[3] There are minor differences in the descriptions of the instruments used by Ptolemy, Copernicus and Brahe, but the basic construction is the same. It is composed of two rods: a vertical pole CD and a rod AB, which pivots about C (see Figure 1.3). AB has plates attached at either end such that the opening in the plate at B is large enough to accommodate the full moon when viewed through the smaller opening in the plate at A.[4] DF is a thin rod which pivots about D. When the moon is sighted in the direction AB, the thin rod can be marked at E, where $CE = CD$. DE is then equal to the chord of $\angle ACD$, the zenith distance of the moon. Ptolemy states that his parallactic instrument is fixed such that the rod AB moves only in the plane of the meridian. Thus the lunar zenith distance must be observed when the moon culminates. In the Copernican instrument, the rod CD is allowed to pivot about a fixed vertical pole.

Ptolemy divides CD into 60 equal parts and the length of DE is read off this scale when the rod is raised to the vertical. This introduces further error

---

[3] Brahe [1], pp. 44-47.

[4] Delambre remarks that the optical parallax resulting from an observation through two holes large enough to accommodate the diameter of the moon at perigee could introduce an error of as much as 10' in altitude, and thus in parallax. Delambre [1], v. 2, pp. 208-209. Tycho Brahe substituted plates with slits rather than holes, which made the instrument more suitable for observing stellar zenith distances. Cf. Brahe [1], p. 46.

into the observation and limits the use of the instrument to zenith distances of less than 60°. The greatest zenith distance of the culminating moon in Alexandria, according to Ptolemy's parameters, is 59;49°. The need to raise DF to the vertical was eliminated by the Arabs, who put the ruler markings on the rod DF.[5] For Copernicus, the length of DF is at least $\sqrt{2} \cdot CD$, and thus zenith distances of up to 90° may be observed and measured. The greatest zenith distance of the culminating moon in Frauenburg, according to Copernicus' parameters, is 82; 47°.

*1.2 The Least Zenith Distance of the Moon and Lunar Latitude (*Almagest *v.12,* De Revolutionibus *iv.15)*

The true zenith distance of the moon cannot be found unless the latitude of the moon is known. However, when the observed zenith distance is approximately equal to the true zenith distance, equation 1.2 can be used to obtain the moon's latitude. The difference between true and apparent positions of the moon is due to the parallax, and lunar parallax will reach a minimum when the moon is at its least distance from the zenith. For any given terrestrial latitude ϕ, there is a specific least zenith distance of the moon. The moon's greatest declination, i.e., its distance from the equator, is the sum of the maximum obliquity of the ecliptic (ϵ) and the inclination of the lunar orbit to the ecliptic (i). For an observer in the northern hemisphere whose latitude is less than the moon's greatest northern declination, the moon will actually pass through the zenith. For a latitude further north, which is the case for both Ptolemy and Copernicus, least zenith distance occurs when the moon culminates at its greatest northern declination, at which time (see Figure 1.4)

$$z = \phi - \epsilon - i. \tag{1.4}$$

The lunar latitude obtained from an observation of least zenith distance,

$$i = \phi - \epsilon - z, \tag{1.5}$$

is the maximum latitude and thus equal to the inclination of the lunar orbit. Once this value is known, the lunar latitude at any given point on the moon's orbit can be found using spherical trigonometry. In order for the conditions of least zenith distance to obtain, the moon must be at the northern limit of its orbit ($w = 0°$) and the ascending node of the lunar orbit must be at the vernal equinox ($\lambda_{\Omega} = 0°$). This configuration implies that the culminating moon is in the summer solstice ($\lambda_m = 90°$).

Through the observation and comparison of lunar eclipses, it is possible to determine when the moon is at the northern limit of its orbit, even though the inclination of the orbit remains unknown. This is discussed by Ptolemy in *Almagest* iv.9, and by Copernicus in *De Revolutionibus* iv.14. By sighting the

[5] Rome [1], p. 69, and Nallino [1], pp. 143-144.

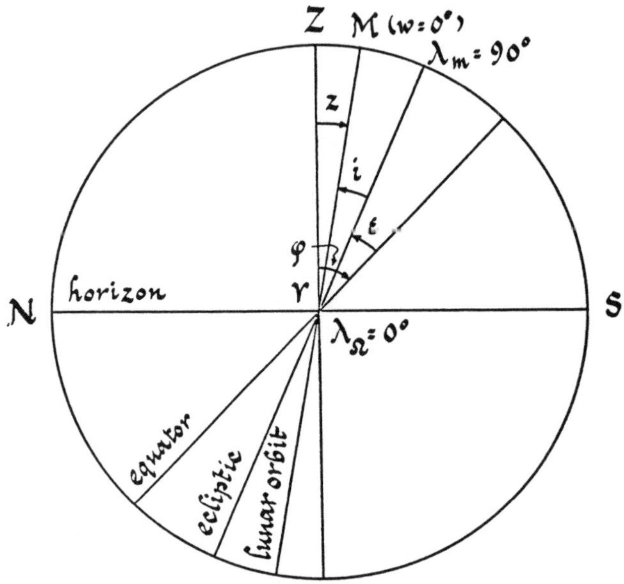

Figure 1.4: Least Lunar Zenith Distance

culminating moon at a time when it is both at the northern limit and at the summer solstice, one obtains the least zenith distance. Ptolemy claims that repeated observations gave a value of 2 1/8° for this quantity.[6].

The maximum latitude is then (equation 1.5)

$$i = 30;58^o - 23;51^o - 2;7^o = 5^o,$$

where 30;58° is Ptolemy's value for the latitude of Alexandria, 23;51° is his value of the obliquity to one sexagesimal place, and 2;7° is 2 1/8° to one sexagesimal place. This value of the latitude agrees with the value given by Hipparchus, which was probably Ptolemy's intent.

Ptolemy does not give any particulars of his observation, but a reasonable date can be reconstructed. The optimal times for observing least zenith distance, when the ascending node is near the vernal equinox, occur once every $18\frac{1}{2}$ years, the period of the revolution of the lunar nodes. This happened in Ptolemy's time in July-August 126 and February-March 145.[7] Since Ptolemy's dated observations in the *Almagest* lie between March 127 and February 141, the date A.D. 126 must be assumed. During the July-August period, there were three occasions when the moon was at the northern limit of latitude.[8]

[6] The parallax under these conditions is indeed quite small. If the lunar distance is taken at its mean value according to Ptolemy, 59^{tr}, then in Figure 1.2, when OC is 1^{tr}, CN is 59^{tr}, and z' is 2 1/8°, the parallax is 0;2,10°. This is below the limits of observation.

[7] P.V. Neugebauer [1].

[8] Tuckerman [1], p. 81.

Britton has determined that the middle of these three was the optimal date for observation, namely August 3, 126. On this day the moon culminated 2 hours before noon, when λ_m was 88° and λ_{Ω} was 0;1°. The other two dates give less satisfactory lunar longitudes.[9]

Copernicus gives the latitude of Frauenburg as 54;19°. Assuming a greatest lunar latitude of 5°, and using Copernicus' value for the obliquity of the ecliptic in his own time, the least zenith distance is (equation 1.4)

$$z = 54;19^o - 23;28^o - 5;0^o = 30;46^o.$$

The lunar parallax at this zenith distance, using Copernicus' mean lunar distance of 60;19^{tr}, would be 0;29,10°, a value too large to be ignored. Copernicus in fact states that the lunar parallax prevented him from determining the lunar latitude experimentally and he merely accepts Ptolemy's value of 5°.

In his commentary on *Almagest* v.12, Reinhold uses Ptolemy's values for the obliquity of the ecliptic and least zenith distance to two sexagesimal places, obtaining a maximum lunar latitude of

$$i = 30;58^o - 23;51,20^o - 2;7,30^o = 4;59,10^o.$$

Figure 1.5 reproduces the diagram in Reinhold's manuscript (Ms.R.173r). There is no commentary by Reinhold on *De Revolutionibus* iv.15.

*1.3 The Greatest Zenith Distance of the Moon and Lunar Parallax (*Almagest *v.12-13,* De Revolutionibus *iv.16)*

If the lunar distance is to be derived from the parallax, it is best to measure the parallax when it is relatively large, since then it is more easily observed and less relative error is introduced. Maximum lunar parallax occurs when the moon is on the horizon, and it is from horizontal parallax that distance is most easily derived. Astronomical refraction, however, prevents an accurate observation of the zenith distance when the moon is close to the horizon. When the observed zenith distance is 90°, the moon is actually 34′ below the horizon,[10] an amount greater than the angular diameter of the moon. Neither Ptolemy nor Copernicus had good control over the effects of refraction and they did not choose to use the horizontal parallax for determining lunar distance. Ptolemy was limited by his parallactic instrament, which measured zenith distance only in the meridian. The closest approach of the culminating moon to the horizon, according to Ptolemy's parameters for Alexandria, is approximately 30°. Copernicus measured lunar parallax in the meridian probably because 1) he followed the example of Ptolemy, 2) this eliminated the need to transform coordinates, and 3) the culminating moon comes within 7° of the horizon in Frauenburg, according to his parameters.

---

[9] Britton [1], p. 159. Cf. also Swerdlow [1], p. 43.

[10] Newcomb [1], p. 29.

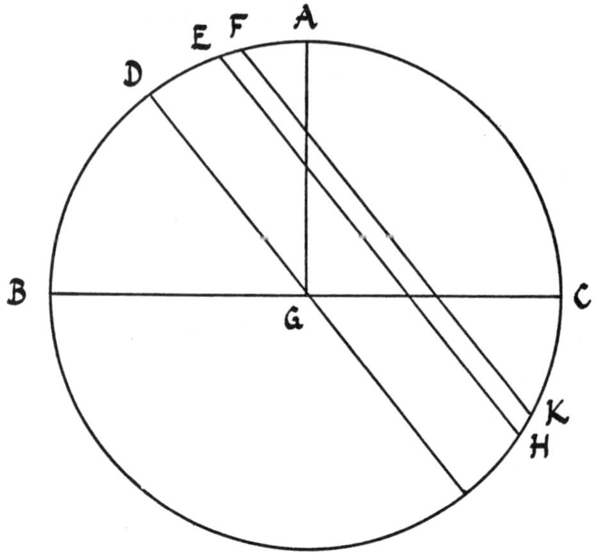

ABC	Meridian of Alexandria	*A*	Zenith
BGC	Horizon		
DG	Equinoxes		
EH	Summer tropic		
FK	Parallel of the moon described in the summer tropic and maximum northern latitude		
EF	Maximum latitude of the moon		

Figure 1.5: Zenith Distance Diagram from Reinhold's Manuscript

Maximum parallax of the culminating moon occurs at greatest lunar zenith distance. For an observer in the northern hemisphere whose co-latitude is less than the moon's greatest declination, the culminating moon will actually appear on the horizon. For a latitude further north, which is the case for Ptolemy and Copernicus, greatest zenith distance occurs when the moon is at its greatest southern declination, when

$$z = \phi + \epsilon + i \tag{1.6}$$

(see Figure 1.6). In order for this configuration to occur, the moon must be at the southern limit of its inclined orbit ($w = 180^o$) and the ascending node must be at the vernal equinox ($\lambda_\Omega = 0^o$). Under these conditions, the culminating moon will be at the winter solstice ($\lambda_m = 270^o$). Using his parameters, Ptolemy would have found a greatest lunar zenith distance of

$$z = 30;58^o + 23;51^o + 5;0^o = 59;49^o,$$

which implies a maximum lunar parallax of 0;50,48° at mean lunar distance. The parameters of Copernicus would produce a greatest lunar zenith distance

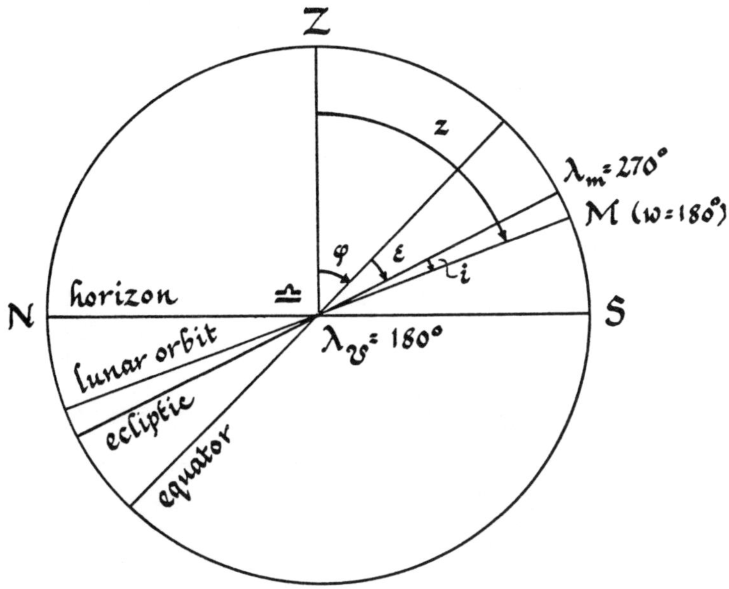

Figure 1.6: Greatest Lunar Zenith Distance

of

$$z = 54;19^o + 23;28^o + 5;0^o = 82;47^o,$$

not far from the horizon. The parallax at this zenith distance, assuming Copernicus' mean lunar distance, would be 0;56,16°.

For both greatest and least zenith distance of the culminating moon, the ascending node is at the vernal equinox, but in one case the moon is at the southern limit of its inclined orbit, and in the other, at the northern limit. Thus the optimal time for observing greatest zenith distance is about half a draconitic month before or after the optimal time for observing least zenith distance. Ptolemy does not do this. He chooses a date nine years later, when the moon is near the *northern* limit of its inclined orbit and the ascending node is at the autumnal equinox. This places the moon in the winter solstice, the only condition Ptolemy specifies for observing greatest zenith distance (see Figure 1.7). At this time, the zenith distance would be, again reconstructing from Ptolemy's parameters,

$$\begin{aligned} z &= \phi + \epsilon - i \\ &= 30;58^o + 23;51^o - 5;0^o = 49;49^o. \end{aligned} \tag{1.7}$$

In Ptolemy's observation, however, the moon was not precisely at the winter solstice nor at its northern limit, so that the result actually obtained by Ptolemy was

$$z = 30;58^o + 23;49^o - 4;59^o = 49;48^o.$$

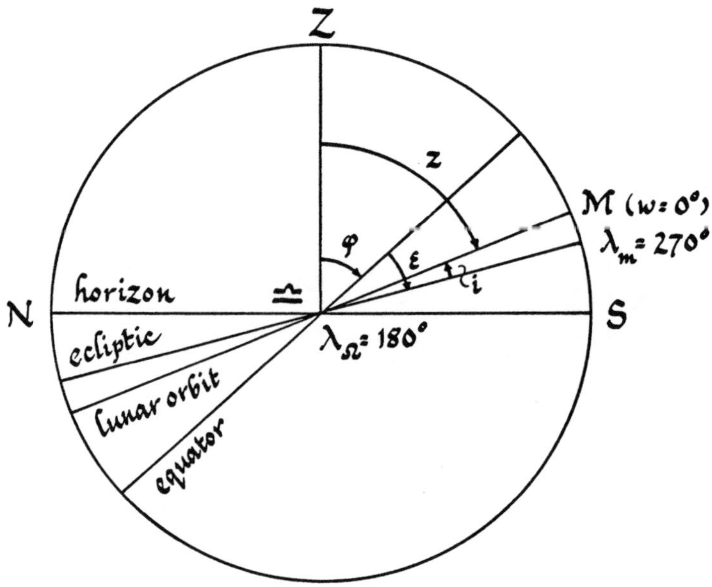

Figure 1.7: Ptolemy's Observation of Lunar Zenith Distance

All dated observations made by Copernicus and reported in *De Revolutionibus* lie between March 1497 and March 1529. The ascending node was at the vernal equinox on July 1, 1522. Copernicus gives two observations of greatest zenith distance. The first is on September 27, 1522, when

$$\begin{aligned} \lambda_m &= 282;32^o, \\ \lambda_\Omega &= 355;24^o, \\ w &= 197;8^o, \end{aligned}$$

and (equation 1.3)

$$z = 54;19^o + 27;41^o = 82;0^o.$$

The second observation was not made near an optimal time but on August 7, 1524, when

$$\begin{aligned} \lambda_m &= 249;39^o, \\ \lambda_\Omega &= 319;22^o, \\ w &= 200;17^o, \end{aligned}$$

and

$$z = 54;19^o + 26;36^o = 80;55^o.$$

It is this second observation which Copernicus chose to use in determining the lunar distance. The details of both observations are described below.

2. THE COMPUTATION OF SOLAR AND LUNAR LONGITUDES

The apparent zenith distance of the moon is obtained from an actual observation, made at a specific time. The true zenith distance, derived from the lunar model, is computed for the time of the observation, a computation requiring the determination of mean and true solar longitude, lunar longitude and latitude, and the declination of the moon. Since the procedures for calculating these quantities are repeated several times in this chapter, they have first been outlined. Tables 1.1 and 1.2 summarize the notation which is used.

2.1 Ptolemaic Solar Longitude

The computation of a solar position from the Ptolemaic model is a simple procedure. There is one mean motion table for the movement of the solar longitude and anomaly. Since the solar apogee has fixed longitude for Ptolemy, the mean solar longitude $\bar{\lambda}_s$ and the solar anomaly $\bar{k}$ always differ by a constant:

$$\bar{k} = \bar{\lambda}_s - 65;30^o.$$

The equation of center c_s is found in the solar correction table as a function of the anomaly. The true solar longitude is then

$$\lambda_s = \bar{\lambda}_s + c_s \qquad \text{for} \qquad 180^o \leq \bar{k} \leq 360^o$$

or

$$\lambda_s = \bar{\lambda}_s - c_s \qquad \text{for} \qquad 0^o \leq \bar{k} \leq 360^o.$$

All Ptolemaic longitudes are measured from the vernal equinox, ♈ 0°.

2.2 Copernican Solar Longitude

The various refinements Copernicus introduced into his solar model make the procedure for computing solar longitude rather lengthy and complicated. Copernicus assigns three motions to the earth: the daily rotation, the annual revolution, and the rotation of the earth's axis (*De Revolutionibus* i.11). The daily rotation accounts for the apparent diurnal motion of the stars, which, according to Copernicus, have no motion whatsoever. The annual revolution accounts for the apparent motion of the sun through the zodiac in the course of a sidereal year. The rotation of the earth's axis was introduced primarily to keep the axis parallel to itself during the course of the earth's annual revolution, but Copernicus makes additional use of this motion. He accounts for the annual precession of the equinoxes by making the period of the rotation of the earth's axis slightly less than a sidereal year.

Copernicus' uncritical acceptance of the accuracy of all observations made by his predecessors led him to believe that the rate of precession was uneven and that the obliquity of the ecliptic varied. He also believed that these two phenomena were related, namely, that the period of irregularity of the obliquity

Table 1.1

SOLAR NOTATION

PTOLEMY

		argument	
$\bar{k}$	mean solar anomaly		
$\bar{\lambda}_s$	mean solar longitude		$\bar{\lambda}_s = \bar{k} + 65;30^o$
c_s	solar equation of center	$\bar{k}$	
λ_s	true solar longitude		$\lambda_s = \bar{\lambda} \pm c_s$

COPERNICUS

$\bar{P}$	360° minus the sidereal longitude of the mean vernal equinox (measured from γ Arietis)		
θ	anomaly of the equinoxes		
c_P	correction of the precession	2θ	
c_ε	correction of the obliquity	θ	
P	360° minus the sidereal longitude of the true vernal equinox measured from γ Arietis)		$P = \bar{P} \pm c_p$
$\bar{k}$	mean solar anomaly		
λ_s^*	"simple" mean solar longitude, i.e., mean sidereal longitude of the sun measured from γ Arietis		
$\bar{\lambda}_s$	mean solar longitude, i.e., mean tropical longitude of the sun measured from the true vernal equinox		$\bar{\lambda}_s = \lambda_s^* + P$
$\bar{l}_s$	"composite" mean solar longitude, i.e., mean tropical longitude of the sun measured from the mean vernal equinox		$\bar{l}_s = \lambda_s^* + \bar{P}$
c_3	solar correction	θ	
k	true solar anomaly		$k = \bar{k} \pm c_s$
c_4	solar correction	k	
c_5	solar correction	k	
c_6	solar correction	θ	
c_s	solar equation of center		$c_s = c_4 + c_5 \cdot c_6$
λ_s	true solar longitude, i.e., true tropical longitude of the sun measured from the true vernal equinox		$\lambda_s = \bar{\lambda}_s \pm c_s$

Table 1.1 (cont.)

PTOLEMY and COPERNICUS

α	right ascension of λ_s
α_o	right ascension of λ_s at epoch
E^o	equation of time in equatorial degrees
	Copernicus: $E^o = \alpha - \alpha_o - (\bar{l}_s(t) - \bar{l}_s(t_o))$
	Ptolemy: $E^o = \alpha - \alpha_o - (\bar{\lambda}_s(t) - \bar{\lambda}_s(t_o))$
E^h	equation of time in hours $E^h = 0;4^{h/o} \cdot E^o$
ϵ	obliquity of the ecliptic
ϕ	terrestrial latitude of the observer
$-^h$	hours (equinoctial)
$-^d$	days
$-^a$	Egyptian year (365^d)
$-^j$	Julian year ($365\ 1/2^d$)
R_s	apparent solar radius
R_m	apparent lunar radius
R_{sh}	apparent shadow radius
D_s	apparent solar diameter
D_m	apparent lunar diameter
S_s	solar distance
S_{ax}	length of the shadow axis
$-^{tr}$	terrestrial radii

Table 1.2

LUNAR NOTATION

		argument	
$\bar{\lambda}_m$	mean lunar longitude		
$\bar{\eta}$	elongation of the mean moon from the mean sun		
$\bar{\gamma}$	mean lunar anomaly		
$\bar{w}$	mean argument of lunar latitude		
c_3	lunar correction	$2\bar{\eta}$	
γ	true lunar anomaly		$\gamma = \bar{\gamma} \pm c_3$
c_4	lunar correction	γ	
c_5	lunar correction	γ	
c_6	lunar correction	$2\bar{\eta}$	
c_m	lunar equation of center		$c_m = c_4 + c_5 \cdot c_6$
λ_m	true lunar longitude		$\lambda_m = \bar{\lambda}_m \pm c_m$
η	elongation of the true moon from the mean sun		$\eta = \bar{\eta} \pm c_m$
w	true argument of lunar latitude		$w = \bar{w} \pm c_m$
β	lunar latitude	w	
δ_m	declination of the moon		
δ_m^*	declination of the projection of the moon on the ecliptic, i.e., declination of the point on the ecliptic which has the same longitude as the moon		
z'	observed zenith distance		
z	true zenith distance		
π	parallax in altitude		$\pi = z' - z$
i	inclination of the lunar orbit		
λ_{Ω}	longitude of the ascending lunar node		
$\lambda_{\mho}$	longitude of the descending lunar node		
S_m	lunar distance		
S/s	ratio of the greatest to least lunar distance		

(3434 years) was twice that of the precession. To account for this, Copernicus introduced two oscillating motions of the polar axis of the earth (*De Revolutionibus* iii.3).[11] The solar model is further modified by a slight sidereal motion of the apsidal line and a small and slow variation in the eccentricity (*De Revolutionibus* iii.20). Al-Battānī had introduced a movement of the line of apsides in his solar model, and Copernicus was familiar with this through Regiomontanus' *Epitome*. However, Copernicus believed not only that these two motions were irregular (again due to an attempt to accomodate the observations of his predecessors), but that their rate of change was connected to the variation in precession and obliquity. The resulting solar model is considerably more complicated than that of Ptolemy, without introducing any essential improvement in the computation of solar longitude.

According to Ptolemaic theory, precession is described as a slow eastward motion of the fixed stars in longitude, which Ptolemy measures from the vernal equinox, while they maintain their latitudes (*Almagest* iii.1). In Copernican theory, the vernal equinox moves at an uneven rate relative to the fixed stars. Copernicus advocated the use of sidereal longitudes measured from γ Arietis, the first zodiacal star listed in Ptolemy's star catalogue (*De Revolutionibus* ii.14).[12] Both the variation in the rate of precession and the use of sidereal longitudes must be kept in mind when working with all Copernican longitudes, particularly in the case of the sun.

The annual mean motion in longitude of the sun or of the planets can be determined for either a tropical or a sidereal year. Copernicus' mean motion tables for the planets are computed for a sidereal year. The addition of the mean motion to the sidereal longitude at epoch produces a sidereal position. The distance of the planet relative to the vernal equinox can then be obtained by adding the precession, the computation of which is described below. For the sun, Copernicus provides tables of the mean motion in longitude for both tropical and sidereal years (*De Revolutionibus* iii.14). He refers to the sidereal motion as the "motus solis aequalis simplicis," or "simple" mean motion of the sun, and to the tropical motion as the "motus solis aequalis compositi," or "composite" mean motion of the sun. The simple solar longitude is measured from γ Arietis, the composite solar longitude is measured from the mean position of the vernal equinox, and at any given moment, the two differ by the sidereal longitude of the mean vernal equinox, otherwise known as the precession. The term 'mean solar longitude,' when used in connection with Copernicus, denotes the mean tropical longitude of the sun measured from the true position of the vernal equinox.

The first step in computing the solar longitude is to obtain the mean mo-

---

[11] Copernicus' mechanism does not in fact produce the motion he assumed (Neugebauer [2], p. 96), nor does it produce results agreeing with the observational data which led him to this device (Dreyer [1], p. 330).

[12] The use of sidereal longitudes was not original with Copernicus. The Canobic Inscription uses Regulus as a zero point, a practice also followed in medieval times. Cf. Neugebauer [1], p. 207.

tions since epoch of the precession of the equinoxes ($\Delta\bar{P}$) and of the anomaly of the equinoxes ($\Delta\theta$) from the tables in *De Revolutionibus* iii.6. P, the sidereal longitude of the true vernal equinox, is

$$P = \bar{P} + c_p \qquad \text{for} \qquad 180^o \leq 2\theta \leq 360^o$$

or

$$P = \bar{P} - c_p \qquad \text{for} \qquad 0^o \leq 2\theta \leq 180^o.$$

In order to compute the corrected, as opposed to mean, obliquity of the ecliptic at a given time, the correction of the obliquity, c_ϵ, is found in the second column of the correction table as a function of θ.

Copernicus gives three tables for the mean motion of the sun (*De Revolutionibus* iii.14). The first two are for the mean solar motion in longitude, and the third is for the mean motion of the solar anomaly. The annual motion of the "simple" sun is computed for a sidereal year, and the simple mean solar longitude, λ_s^*, measures the mean sidereal longitude of the sun from γ Arietis. The motion of the "composite" sun is computed for a mean tropical year, and the composite mean solar longitude, $\bar{l}_s$, measures the mean tropical longitude of the sun from the mean vernal equinox. The two are related as follows:

$$\bar{l}_s = \lambda_s^* + \bar{P}.$$

The mean longitude of the sun measured from the true vernal equinox, $\bar{\lambda}_s$, may be found from either the simple or composite solar longitude:

$$\bar{\lambda}_s = \lambda_s^* + P$$

or

$$\bar{\lambda}_s = \bar{l}_s \pm c_p.$$

In computing a solar position, Copernicus usually states only the mean longitude of the sun, $\bar{\lambda}_s$, from which it is not clear whether he has used λ_s^* or $\bar{l}_s$.

Since Copernicus' solar apogee is not sidereally fixed, the mean solar anomaly, $\bar{k}$, must be computed separately using the third mean motion table. In the solar correction table, c_3 and c_6 are found as functions of θ, the anomaly of the equinoxes. The true solar anomaly k is

$$k = \bar{k} + c_3 \qquad \text{for} \qquad 0^o \leq \theta \leq 180^o$$

or

$$k = \bar{k} - c_3 \qquad \text{for} \qquad 180^o \leq \theta \leq 360^o.$$

The solar corrections c_4 and c_5 are then found as functions of k. The equation of center for the sun is

$$c_s = c_4 + c_5 \cdot c_6\ .$$

The longitude of the true sun λ_s, measured from the true vernal equinox, is

$$\lambda_s = \bar{\lambda}_s + c_s \qquad \text{for} \qquad 180^o \leq k \leq 360^o$$

or

$$\lambda_s = \bar{\lambda}_s - c_s \qquad \text{for} \qquad 0^o \leq k \leq 180^o$$

In terms of λ_s^* and $\bar{l}_s$,

$$\lambda_s = \lambda_s^* + P \pm c_s$$

and

$$\lambda_s = \bar{l}_s \pm c_p \pm c_s.$$

2.3 Ptolemaic Lunar Longitude and Latitude

Ptolemy has four mean motion tables for the moon which, together with the appropriate epoch values, produce the mean lunar longitude $\bar{\lambda}_m$, the mean lunar anomaly $\bar{\gamma}$, the mean argument of lunar latitude $\bar{w}$, and the elongation of the mean moon from the mean sun $\bar{\eta}$. In determining the lunar longitude for a given date, Ptolemy does not always use his table for $\Delta\bar{\lambda}_m$. The elongation, which is used in computing lunar corrections, can be combined with the mean solar longitude to produce the mean lunar longitude:[13]

$$\bar{\lambda}_m = \bar{\lambda}_s + \bar{\eta}.$$

The calculation of the lunar equation of center for Ptolemy's complex lunar model is similar to the calculation of Copernicus' solar equation of center. The lunar corrections c_3 and c_6 are found as functions of $2\bar{\eta}$. The true lunar anomaly is

$$\gamma = \bar{\gamma} + c_3 \qquad \text{for} \qquad 0^o \leq 2\eta \leq 180^o$$

or

$$\gamma = \bar{\gamma} - c_3 \qquad \text{for} \qquad 180^o \leq 2\eta \leq 360^o.$$

c_4 and c_5 are then found as functions of γ. The equation of center for the moon is

$$c_m = c_4 + c_5 \cdot c_6.$$

The true lunar longitude is

$$\lambda_m = \bar{\lambda}_m + c_m \qquad \text{for} \qquad 180^o \leq \gamma \leq 360^o$$

or

$$\lambda_m = \bar{\lambda}_m - c_m \qquad \text{for} \qquad 0^o \leq \gamma \leq 180^o.$$

The true argument of latitude is

$$w = \bar{w} \pm c_m.$$

The lunar latitude β is found in the last column of the correction table as a function of w.

[13] A problem arises here in that Ptolemy may combine $\bar{\lambda}_s$ computed for the apparent time with $\bar{\eta}$ computed for the corrected time. The resulting lunar longitude may not agree with a more accurately computed value, obtained from the mean motion tables for the corrected time. This appears to be the case in Ptolemy's parallax observation described below.

2.4 Reinhold's Recomputation of Ptolemaic Lunar Longitude and Latitude

When Reinhold recomputes Ptolemy's lunar longitude, he determines the mean motion of $\bar{\lambda}_m$, rather than $\bar{\eta}$. The elongation is then found from

$$\bar{\eta} = \bar{\lambda}_m - \bar{\lambda}_s.$$

All longitudes mentioned by Reinhold in his commentary on the *Almagest* are, of course, measured from the vernal equinox, ♈ 0°.

2.5 Copernican Lunar Longitude and Latitude

Copernicus gives only three mean motion tables for the moon (*De Revolutionibus* iv.4): motion in elongation, anomaly and argument of latitude. The mean elongation ($\bar{\eta}$) measures the distance of the mean lunar longitude ($\bar{\lambda}_m$) from the mean solar longitude ($\bar{\lambda}_s$):

$$\bar{\lambda}_m = \bar{\lambda}_s + \bar{\eta}.$$

Thus $\bar{\lambda}_m$ is measured from the true vernal equinox and cannot be obtained from Copernicus' tables without prior knowledge of the mean solar longitude. The procedure for computing lunar longitude and latitude is identical to Ptolemy's, with one exception. Copernicus finds the true elongation of the moon (η) from the mean solar longitude

$$\eta = \bar{\eta} \pm c_m$$

and combines this with the mean solar longitude to obtain the true lunar longitude

$$\lambda_m = \bar{\lambda}_s + \eta.$$

With this method, the mean solar longitude should be computed for the corrected time.

2.6 Reinhold's Recomputation of Copernican Solar and Lunar Longitude

For the two observations in *De Revolutionibus* iv.16, Reinhold uses the simple sun to compute the solar longitude, though elsewhere he uses the composite sun. He does not use the mean solar longitude to obtain the true solar longitude. Instead he finds $\lambda_s^* \pm c_s$ and combines this with the true precession:

$$\lambda_s = (\lambda_s^* \pm c_s) + P.$$

Reinhold does not find the mean solar longitude for the corrected time, even though he uses it in the equation

$$\lambda_m = \bar{\lambda}_s + \eta.$$

One other difference between the methods of Copernicus and Reinhold is that Reinhold has tabulated an additional function to facilitate the computation of the true obliquity. For Copernicus, the correction of the obliquity is

$$c_\epsilon(\theta) = 0;24^o \cdot c_2(\theta),$$

where θ is the anomaly of the equinoxes, $c_2(\theta)$ is the correction function found in the second column (after the arguments) of Copernicus' correction table for precession and obliquity (*De Revolutionibus* iii.8), and 0;24° is the difference between Copernicus' values of the maximum and minimum obliquity of the ecliptic. The resulting correction c_ϵ is always added to the minimum obliquity to obtain the true obliquity. Reinhold has simplified this procedure by computing an additional function:

$$c_3(\theta) = 0;24^o \cdot c_2(\theta) - 0;12^o.$$

This correction, found in the third column (not counting the difference columns) of Reinhold's correction table (*Prutenic Tables*, f.16v-19r), is directly added to or subtracted from the mean obliquity.

2.7 *The Equation of Time*

For Ptolemy, the equation of time in equatorial degrees is

$$E^o = \alpha - \alpha_o - (\bar{\lambda}_s(t) - \bar{\lambda}_s(t_o)).$$

Reinhold always uses the composite sun, rather than the mean sun, in his equation of time:

$$E^o = \alpha - \alpha_o - (\bar{l}_s(t) - \bar{l}_s(t_o)).$$

This is reasonable, since the composite sun measures a uniform motion $(\lambda_s^* + \bar{P})$, whereas Copernicus' mean sun varies with the precession $(\lambda_s^* + \bar{P} \pm c_p)$. I assume Copernicus also used the composite sun in computing his equation of time. The conversion from degrees to hours is

$$E^h = 0;4^{h/o} \cdot E^o.$$

The epoch position of the composite sun and the right ascension of the true solar longitude at epoch are calculated in Appendix A.

3. THE OBSERVATIONS

The observations used by Ptolemy and Copernicus to determine lunar parallax were not made at the optimal times for observing greatest lunar zenith distance in the meridian ($w = 180^o$, $\lambda_\Omega = 0^o$, $\lambda_m = 270^o$). In Ptolemy's observation, the moon was at the northern limit of its orbit ($w = 0^o$). In the Copernican observation from which the lunar distance is ultimately derived, the moon is 20° from the winter solstice, 20° from the southern limit of the orbit, and the ascending node is 40° from the vernal equinox. Thus I assume that Ptolemy and Copernicus exercised some control over their choice of a parallax observation and consequently over their final results for true lunar distance.

The relative position of the sun and moon is not among the conditions which produce greatest lunar zenith distance in the meridian, yet both Ptolemy and Copernicus present observations made when the moon was near quadrature. In the case of Copernicus, this is quite understandable. His lunar model was designed to correct the close approach of the moon to the earth at quadrature produced by the Ptolemaic model. Copernicus' parallax observation yields a lunar distance in quadrature much greater than Ptolemy's, but a mean distance close to Ptolemy's value of the mean distance in syzygy. Ptolemy never refers explicitly to the great variation in distance and apparent lunar diameter which would be produced by his model. One might expect that he would prefer not to call attention to it. His parallax observation, however, actually confirms the close approach of the moon to the earth in quadrature. This confirmation may in fact have been Ptolemy's aim, as suggested by Swerdlow.[14] On the other hand, Ptolemy may have chosen an observation in quadrature because parallax is greater, and thus easier to observe, when the moon is closer to the earth.

*3.1 Ptolemy's Observation (*Almagest *v.13)*

Ptolemy's parallax observation was made on October 1, 135, 5;50^h after noon, when the moon was culminating and the sun was about to set. He found that the distance of the moon from the zenith of Alexandria was 50;55°.[15] The time since epoch (Era Nabonassar) was 882^a 72^d 5;50^h.

Ptolemy's results are listed in the first column of Table 1.3 and the computations are described in detail in Appendix A. The eqution of time, accurately computed, should be –0;27,28^h, but it is reasonable for Ptolemy to round this to half an hour. Since the mean solar longitude is computed for the corrected time of 5;20^h, it is not consistent to find

$$\begin{aligned}\bar{\lambda}_m &= \bar{\lambda}_s + \bar{\eta} \\ &= 187;31^o + 78;13^o = 265;44^o,\end{aligned}$$

as Ptolemy does. Using $\bar{\lambda}_s$ for the corrected time, or simply using Ptolemy's mean motion table for $\bar{\lambda}_m$, one obtains 265;43° for the mean lunar longitude.

In this observation, the moon is near the winter solstice but not at its greatest southern declination. In fact it is very close to the northern limit ($w = 2;60^o$). In Figure 1.8, AB represents a 90° arc of the ecliptic, BC is a 90°

---

[14] Swerdlow [1], pp. 50-51.

[15] Ptolemy's report of the zenith distance is unusual in that he gives a detail from the actual observation, namely, that the length of DE (see Figure 1.3) was 51;35. The angle whose chord is 51;35, according to Ptolemy's table of chords, is 50;55,3,0°. Delambre suggests that since the length of DE is given as 51 + 1/2 + 1/12, the scale on CD was probably divided into twelfths, or 5' intervals. Delambre [1], v. 2, p. 209.

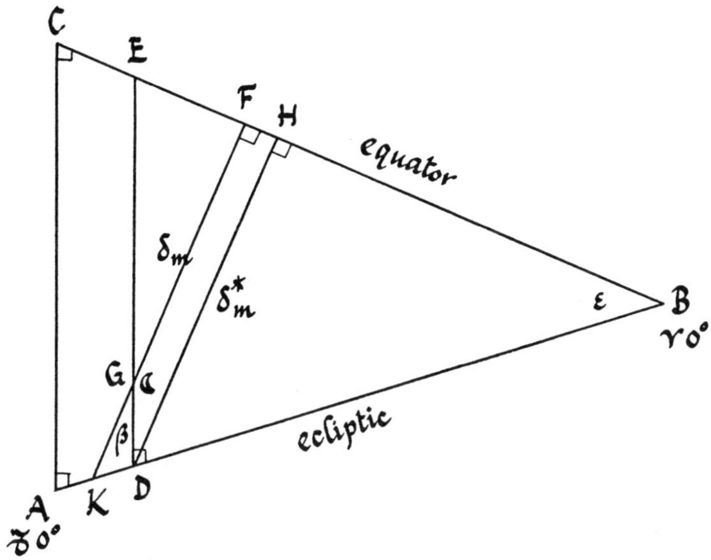

Figure 1.8: Ptolemy's Lunar Parallax Observation

arc of the equator, B is the vernal equinox, A is the winter solstice, and the moon is at G. Ptolemy computes the declination of the point on the ecliptic which has the same longitude as the moon:

$$DH = \delta_m^*(273;10^o) = 23;49^o,$$

In order to find the true zenith distance, he must then combine this declination with the latitude, DG (equation 1.2):

$$z = 30;58^o + 23;49^o - 4;59^o = 49;48^o.$$

The parallax of the moon is thus (equation 1.1)

$$\pi = 50;55^o - 49;48^o = 1;7^o.$$

3.1.1 Reinhold's Recomputation of Ptolemy's Observation

In a preliminary discussion of this observation (Ms.R.173b), Reinhold says that since Ptolemy used the parallax of the moon to find the lunar distance, he chose a lunar position in the meridian circle when the moon had maximum parallax. In particular, Reinhold specifies that the moon should be at the winter solstice with greatest southern latitude. Nowhere does he mention Ptolemy's actual use of the northern latitude. He adds that the moon should be in quadrature with respect to the sun, since this is the time of the moon's nearest approach to the earth, and thus when parallax, as a function of distance, will be greatest.

Table 1.3

PTOLEMY'S LUNAR PARALLAX OBSERVATION

	PTOLEMY	REINHOLD's Rccomputation
apparent time since epoch	$882^a\ 72^d\ 5;50^h$	$882^a\ 72^d\ 5;50^h$
corrected time since epoch	$882^a\ 72^d\ 5;20^h$	$882^a\ 72^d\ 5;20^h$
z'	50;55°	50;55°
$\bar{k}$	[122; 1°]†	122; 1,13,46°
$\bar{\lambda}_s$	187;31°	187;31,14°
c_s	[2; 3°]	2; 3°
λ_s	185;28°	185;28°
$\bar{\lambda}_m$	265;44°	265;43,14,52°
$\bar{\eta}$	78;13°	78;13, 0°
$\bar{\gamma}$	262;20°	262;19,51,15°
$\bar{w}$	354;40°	354;40,12,32°
$2\bar{\eta}$	[156;26°]	156;26, 0°
c_3	[7;40°]	7;40°
γ	[270; 0°]	269;59,51°
c_4	[4;59°]	4;59, 0°
c_5	[2;35°]	2;35°
c_6	[0;56,42°]	0;56,42°
$c_5 \cdot c_6$	[2;26,28°]	2;26,29°
c_m	7;26	7;25,29°
λ_m	273;10°	273; 8,44°
w	2; 6°	2; 5,22°
β	4;59°	4;59°
δ^*_m	23;49°	23;49,23°
z	49;48°	49;48°
π	1; 7°	1; 7°

† Numbers in brackets do not appear explicitly in the text, but are reconstructed.

Ptolemy's calculation of the parallax was carefully done. Reinhold's recomputation reflects accuracy to more sexagesimal places but does not produce a different result. His values are listed in the second column of Table 1.3, where they can be compared with Ptolemy's. Reinhold's commentary on the *Almagest* contains the values for all intermediate steps in the computation, e.g., $c_3, c_4, c_5, c_6, c_5 \cdot c_6$, and the individual values of the mean motion in 810^a, 72^a, etc.

Reinhold uses Ptolemy's mean motion tables to compute the mean lunar longitude, but not to compute the elongation. His value of the elongation indicates that he must have recomputed the mean solar longitude for the corrected time. He does not recompute the equation of time for this observation, but merely states Ptolemy's correction. This is unusual in that for all his recomputed Copernican observations, and for the two other Ptolemaic observations which he discusses in his commentary on the *Almagest*, Reinhold computes the equation of time himself. There is only one obvious error in Reinhold's calculations. In the addition of $\bar{w}$ and c_m to obtain w, he writes $2;5,\underline{2}2^o$ instead of $2;5,\underline{4}2^o$. This does not affect the value of the latitude, however.

In discussing the declination of the moon, Reinhold explicitly states that he finds the distance between the ecliptic and the equator on a great circle through the poles of the ecliptic and through the point on the ecliptic which has the same longitude as the moon (DE in Figure 1.8). Thus he does not use the lunar declination of the point on the ecliptic ($\delta^*_m = DH$), as Ptolemy does, nor the lunar declination ($\delta_m = GF$). The question of which arc to use for the lunar declination occurs in zenith distance observations where the longitude of the moon is not exactly that of the summer or winter solstice. Reinhold's discussion of this problem is quoted in full below. I have added the letters in brackets which refer to Figure 1.8.

> The arc of the great circle passing through the poles of the ecliptic and ♑ 3;9°, included between the equator and the ecliptic, is 23;49,23° [DE]. This differs very little from the arc of the great circle through the poles of the equator and the same position in ♑, included between the equator and the ecliptic [$DH = \delta^*_m$], since the position of the eclipctic held by the moon in this case is near the tropic, where the circle would pass through the poles of both the circles. Therefore, the arc of latitude of the moon can be considered without error as the arc of the great circle passing through the poles of the equator and the position of the moon, included between the inclined circle of the moon and the ecliptic [GK], although the latitude of the moon in the strict sense is the arc of the great circle through the poles of the ecliptic and the position of the moon similarly intercepted [DG].
> (Ms.R.174r)

The determination of true zenith distance requires the combination of the lunar latitude, measured perpendicular to the ecliptic, the lunar declination, measured perpendicular to the equator, and the terrestrial latitude, measured perpendicular to the horizon. I assume Reinhold found DE, perpendicular to the ecliptic, because it makes more sense to combine this quantity with the lunar latitude, DG, than to combine the declination DH with the latitude DG, as Ptolemy does. Reinhold's value of DE follows from a lunar longitude of 273;10°, Ptolemy's value. A lunar longitude of 273;8,44°, Reinhold's value, or the rounded value 273;9°, will not produce 23;49,23° for DE. Reinhold's discussion of the declination ends with the remark "Haec prolixe declarat et demonstrat commentarium theoriis hoc loco".

At the end of his computations, Reinhold calculates the length of daylight in Alexandria at the time of the observation, obtaining 177;30°.[16] The half-length of daylight is thus 88;45°, and dividing by $15^{o/h}$, the end of daylight occurs at $5;55^h$. Since the time of the observation was $5;50^h$, this confirms Ptolemy's statement that the sun was setting.

In the final computation, Reinhold finds the parallax of the moon only to the nearest minute, and his result agrees with Ptolemy's. Figure 1.9 is reproduced from Reinhold's manuscript (Ms.R.174v). The meridian circle of Alexandria is represented by $\alpha\beta\gamma\delta$, $\gamma\epsilon\delta$ is the horizon, α is the zenith, $\beta\epsilon$ is the equator, $\zeta\epsilon$ is the ecliptic, and η represents the true position of the moon as it culminates. The true zenith distance is then

$$\begin{aligned} \alpha\eta &= \alpha\beta + \beta\zeta - \zeta\eta \\ &= 30;58^o + 23;49^o - 4;59^o = 49;48^o. \end{aligned}$$

Since the observed zenith distance was 50;55°, the lunar parallax is 1;7°.

At this point in the commentary Reinhold makes some remarks which are intended to elucidate Ptolemy's procedure. He says that from this parallax (1;7°), Ptolemy calculated the following:

> 1. The distance of the moon from the earth when the moon was in quadrature
> 2. The distance of the moon at the apogee and perigee of the eccentric
> 3. From this hypothesis he wrote or prepared a table of general parallax. (Ms.R.174v)

3.1.2 Copernicus' Summary of Ptolemy's Observation

At the beginning of *De Revolutionibus* iv.16, Copernicus reviews Ptolemy's calculation of lunar parallax and distance. He merely states the values given by Ptolemy, omitting $\bar{\lambda}_s$ and $\bar{\lambda}_m$. With the exception of the true lunar longitude, the values are identical to those found in the *Almagest*. Copernicus has 273;9° for λ_m, rather than 273;10°. The value 273;9° is also found in the *Epitome*.[17]

[16] This value is a little large, since, from Ptolemy's table in *Almagest* ii.8, a solar longitude of 185;28° at the latitude of the low countries of Egypt, 30;22°, corresponds to a length of daylight of 177;25°. The value for Alexandria would be a little smaller. Reinhold may have been approximating the value, though this is unlike him.

[17] Regiomontanus [1], v. 15.

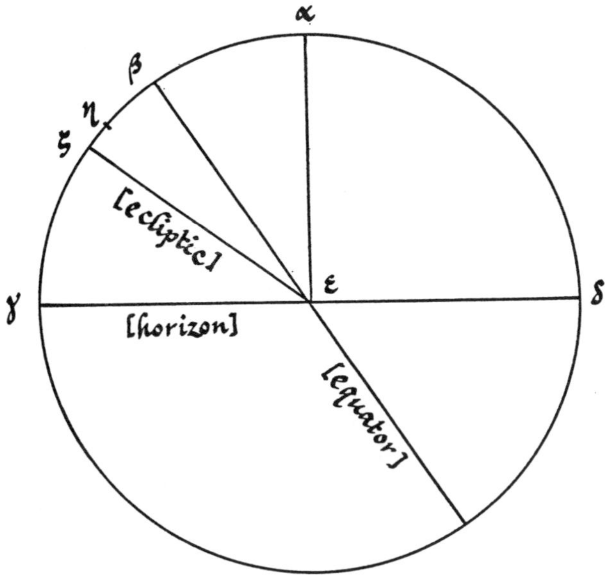

Figure 1.9: Reinhold's Illustration of Ptolemy's Lunar Parallax Observation

Thus Ptolemy's error in using the uncorrected time for the mean solar longitude and the corrected time for the mean lunar elongation has been rectified somewhere between his own time and that of Regiomontanus.

*3.2 Copernicus' Observations (*De Revolutionibus *iv.16)*

Two problems arise in attempting to recompute and verify the results of Copernicus' observations. The manuscript of *De Revolutionibus* indicates that Copernicus originally found the mean motions of the lunar elongation, the lunar anomaly, and the argument of latitude and computed tables accordingly. Subsequently, however, he corrected his values and made appropriate changes in the tables. In an attempt to determine which values Copernicus used in calculating his observations, I have computed the mean motions for the different values. The subscripts a, b, and c refer to the first, second, and third values of $\bar{\eta}$ and $\bar{\gamma}$, respectively. There are only two values of $\bar{w}$. The specific values, as well as the actual computations, are found in Appendix A.

The second problem also concerns changes made in the manuscript. The mean motions, corrections, final parallax, even the time and apparent zenith distance of the observation itself have been altered by Copernicus. Some of these changes can be accounted for reasonably, others cannot. I have used the subscripts 1, 2, and 3 to indicate the original and subsequent values of quantities in the manuscript. These changes are summarized in Tables 1.4 and 1.8.

3.2.1 Copernicus' First Observation

Ignoring for the moment the time of day, Copernicus' first observation was made on September 27, 1522. The zenith distance of the moon, as it crossed the meridian in Frauenburg, was 82;50°. Copernicus first gives the time of day as $5\frac{2}{3}$ equinoctial hours after noon, near the time of sunset. The fractional part of the hours has been changed in the manuscript to $1/3^h$, and subsequently has been changed back to $2/3^h$. The time is given again when Copernicus states the time elapsed since epoch (Era Christ), first written as 1522 Egyptian years, 284^d, 17 $1/3^h$. The $1/3^h$, however, has been changed to $2/3^h$, and this is the value I have assumed below in recomputing Copernicus' results.

Of the many quantities which must be computed to obtain the true zenith distance, Copernicus gives a select few. In particular, he gives only the final result of his solar longitude computation. The values found by Copernicus for this observation are listed in Table 1.4. The first column of numbers contains those values which were written first in the *De Revolutionibus* manuscript. The second column indicates corrections made by Copernicus in the manuscript. The values appearing in the first printed edition (Nuremberg) are listed in the last column.

Copernicus has two values for the true solar longitude: first 193;34°, and then 193;29°. Since there are so many small steps in the computation of solar longitude, and thus many places where error could be introduced, it is difficult to account for either value. An accurate recomputation from Copernicus' tables for the apparent time of the observation yields 193;31°. The correction or change made by Copernicus, from 193;34° to 193;29°, cannot be due to his indecision over the time of the observation, since the mean motion of the composite sun in one third of an hour amounts to less than one minute of arc.

Accurately computed, using my own recomputed value of the solar longitude, the equation of time should be

$$\begin{aligned} E^o &= \alpha(193;31^o) - \alpha_o(279;38^o) - (194;48^o - 278;2^o) \\ &= -4;52^o, \end{aligned}$$

where α_o has been computed for maximum obliquity and α for minimum obliquity. Then

$$E^h = 0;4^{h/o} \cdot (-4;52^o) = -0;19,28^h.$$

Copernicus' final value for E^h is $-0;16^h$.[18]

The values of the mean lunar motions, $\bar{\eta}$, $\bar{\gamma}$, and $\bar{\varpi}$, at the time of the observation, have been summarized in Table 1.5. Copernicus gives only one value for the elongation, and thus his value for the lunar correction c_3 should have been

$$c_3(2\bar{\eta}) = c_3(174;12^o) = 1;1,52^o.$$

---

[18] Copernicus appears to have originally written $17;23^h$ as the corrected time since epoch, implying a value of $-0;17^h$ for E^h. The critical edition reads $17;13^h$ as the original value. Taken in conjunction with the 5 $1/3^h$, this would imply $-0;7^h$ for E^h.

Table 1.4

COPERNICUS' FIRST LUNAR PARALLAX OBSERVATION

	First version in Manuscript	Changes in Manuscript	Nuremberg edition
hours of time of observation	$5 1/3^h$*	$5 2/3^h$	$5 2/3^h$
hours of apparent time since epoch	$17 1/3^h$	$17 2/3^h$	$17 2/3^h$
hours of corrected time since epoch	$17;23^h$	$17;24^h$	$17;24^h$
z'	82;50°		82;50°
λ_s	193;34°	193;29°	193;29°
$\bar{\eta}$	87; 6°		87; 6°
$\bar{\gamma}$	357;39°		357;39°
γ	358;29°	358;40°	358;40°
c_m	0; 7°		0; 7°
λ_m	282;37°	282;32°	→ 282;33°
$\bar{w}$	194;49°†	197; 1°	197; 1°
w	194;56°	197; 8°	197; 8°
β	4;50°	4;47°	4;47°
δ_m	27;41°		27;41°
ϕ_F	54;19°		54;19°
z	82°		82°
π	0;50°		0;50°

* Changed from $5 2/3^h$.
† Changed from 199;49°.

From his values of γ and $\bar{\gamma}$, however, it is obvious that he used 1;1° for c_3. This can be accounted for if one assumes 1;1,52° was truncated.

Copernicus states that the lunar equation of center c_m is 0;7°, and this is indeed the value he uses. It does not follow from his value of γ however. From Copernicus' own correction table,

$$c_4(358;40^o) = 0;6,13^o,$$

$$c_5(358;40^o) = 0;3,7^o,$$

and

$$c_6(174;12^o) = 1;0^o.$$

Thus

$$c_m = c_4 + c_5 \cdot c_6 = 0;9,20^o.$$

A similar problem arises in Reinhold's recalculation of the lunar equation.

The change of 282;37° for the true lunar longitude to 282;32° corresponds to the change of the true solar longitude from 193;34° to 193;29°.

The two values for the lunar longitude can be explained as follows:

$$\begin{aligned}\lambda_m &= \lambda_s + c_s + \bar{\eta} + c_m \\ &= 193;33^o + 1;51^o + 87;6^o + 0;7^o = 282;37^o,\end{aligned}$$

where 193;33° is the true solar longitude corrected by the mean motion of the sun in –0;16^h (Copernicus' value for the equation of time), 1;51° is my recomputed value of the solar correction, and 87;6° and 0;7° are Copernicus' values for the mean elongation and lunar correction, and

$$\lambda_m = 193;28^o + 1;51^o + 87;6^o + 0;7^o = 282;32^o.$$

It is much more likely, however, that Copernicus did not correct the solar longitude for the equation of time, and used the truncated value 1;50° for the solar equation, arriving at the same results.

Copernicus' two values for the true argument of latitude correspond to his last two values of the mean argument of latitude and a lunar correction of 0;7°:

$$w = \bar{w} + c_m$$

$$w_1 = 194;49^o + 0;7^o = 194;56^o$$

$$w_2 = 197;1^o + 0;7^o = 197;8^o.$$

Using Copernicus' lunar correction table, one finds in the last column

$$\beta(194;56^o) = 4;49,56^o.$$

Copernicus' original manuscript value of β is 4;50°. Using his second value of w,

$$\beta(197;8^o) = 4;46,27^o.$$

Table 1.5

MEAN LUNAR MOTIONS – COPERNICUS AND REINHOLD

	Mean motions at the time of the observation computed from Copernicus' various mean values			Values of the mean notions at the time of the observation in Copernicus' Manuscript and the printed edition (N)			Reinhold
First Observation (September 27, 1522)							
	a	b	c	1	2	N	
$\bar{\eta}$	87; 4,30°	87; 6, 2°	87; 6, 6°	87; 6°		87; 6°	87; 4,38°
$\bar{\gamma}$	357;39, 7°	357;42,46°	357;40, 1°	357;39°		357;39°	357;22,40°
$\bar{\varpi}$	196;41,51°	197; 1,24°		194;49°*	197; 1°	197; 1°	197; 2,32°
Second Observation (August 7, 1525)							
	a	b	c	1	2	N	
$\bar{\eta}$	97; 5,17°	97; 6,49°	97; 6,53°	97; 5°		→ 97; 6°	97; 5,11°
$\bar{\gamma}$	242;10, 3°	242;13,43°	242;10,57°	242;10°		242;10°	241;53,18°
$\bar{\varpi}$	192;59, 8°	193;18,43°		193;19°		193;19°	193;19,33°

* Changed from 199;49°.

Copernicus' second manuscript value of the latitude is 4;47°. If the latitude is computed accurately, rather than from the table, the result agrees with Copernicus:

$$\beta = \sin^{-1}(\sin 5^o \cdot \sin(270^o - 197;8^o)) = 4;47^o.$$

The agreement may be a fortuitous coincidence.

Copernicus has only one value for the lunar declination, 27;41°. It is not clear whether he found the declination of the true lunar position (δ_m) or the sum of the latitude and the declination of the point on the ecliptic corresponding to the lunar longitude ($\beta + \delta^*_m$). In the first case, using Copernicus' original lunar longitude and latitude, the declination is (see Figure 1.10a)

$$\delta_m(282;37^o, 4;50^o) = 27;41^o,$$

and using the corrected longitude and latitude

$$\delta_m(282;32^o, 4;47) = 27;38^o.$$

In the second case, the declination is either

$$\delta^*_m(282;37^o) + 4;50^o = 27;41,55^o,$$

or

$$\delta^*_m(282;32^o) + 4;47^o = 27;39,20^o.$$

Even though Copernicus' result can be obtained using the lunar declination δ_m, which requires the solution of several spherical triangles (see Appendix A), this may not have been his method. In the second observation it appears that he found the declination of the point on the ecliptic which has the same longitude as the moon, a simpler procedure. He may have done the same here, only incorrectly. In any event, one thing is clear regardless of which method he used. Copernicus originally computed the entire observation, down to the value of the parallax, obtaining essentially those values in the first column of Table 1.4. He later went back and made corrections, which appear in the second column. He did not bother, however, to recompute the lunar declination, which is thus in error by three mintues, as is the parallax.

Finally, since the latitude of Frauenburg, according to Copernicus, is 54;19°, the true zenith distance is

$$\begin{aligned} z &= \phi + \delta_m \\ &= 54;19^o + 27;41^o = 82;0^o. \end{aligned}$$

The observed zenith distance was 82;50°, and thus the lunar parallax is

$$\pi = z' - z = 0;50^o.$$

At this point Copernicus says that according to Ptolemy's teaching the parallax should have been 1;17°. An accurate recomputation with Ptolemy's parallax table yields 1;18°. Copernicus' own parallax table gives 0;49,57° for π. Copernicus' intention, in part, is to show that his lunar theory produces smaller parallaxes than Ptolemy's.

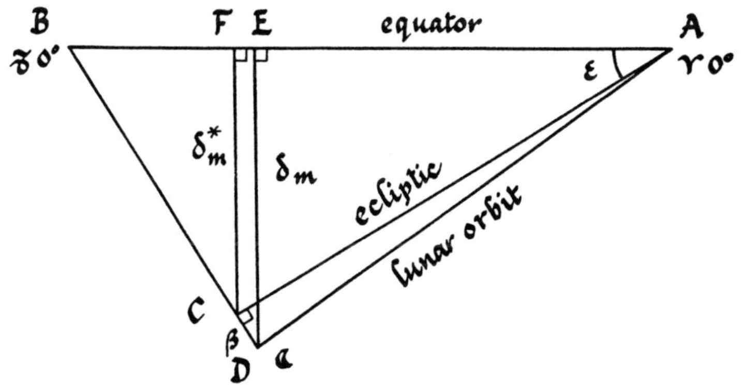

Figure 1.10a: Copernicus' First Lunar Parallax Observations

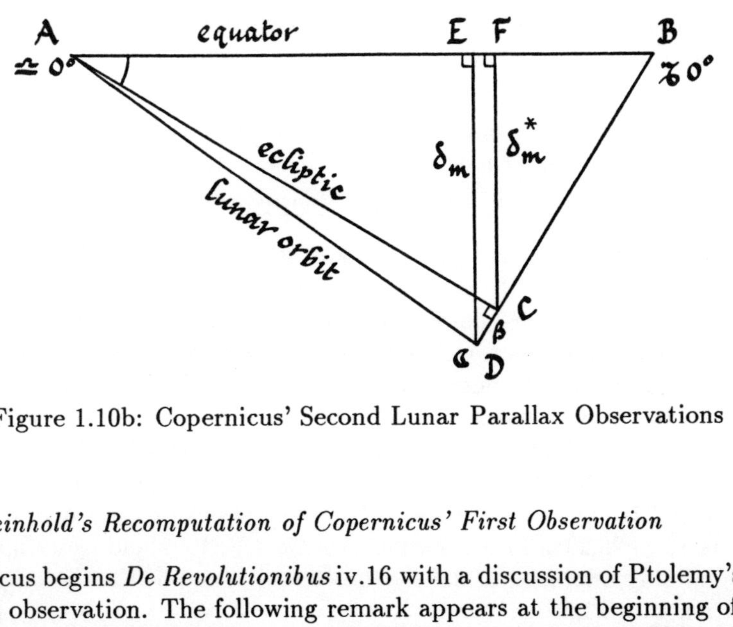

Figure 1.10b: Copernicus' Second Lunar Parallax Observations

3.2.2 Reinhold's Recomputation of Copernicus' First Observation

Copernicus begins *De Revolutionibus* iv.16 with a discussion of Ptolemy's lunar parallax observation. The following remark appears at the beginning of Reinhold's commentary on this chapter: "Ptolemaei supputationem alibi scripsi pro M. Johanne Aurifabro" (Ms.R.50v). This is the only reference in the Copernican commentary to Reinhold's commentary on the *Almagest.*

Following this, Reinhold gives the time of Copernicus' first observation as the 27th of September, 1522 A.D., 5;40^h after noon. The sun was setting in Frauenburg and the center of the moon, observed in the meridian circle, was 82;50° from the zenith. The apparent time since epoch was thus 1522 Egyptian years, 284 days, 17;40^h. In the margin Reinhold has written 269 days. This is the number of days in excess of the 1521 Julian years since epoch. In the manuscript, Reinhold computes mean motions for Julian years.

The values used by Reinhold in computing the solar longitude and equation of time are listed in Table 1.6. They are correct, i.e., they agree with my own recomputed values, to the number of places shown. The computations themselves are discussed in detail in Appendix A. Table 1.7 lists the results for the computation of lunar longitude, latitude and parallax. The first column of numbers contains those values given by Copernicus in the Nuremberg edition of *De Revolutionibus.* Reinhold's mean lunar motions can be compared with those of Copernicus in Table 1.5. If Reinhold's redetermination of the mean lunar motions can be considered an improvement, then Copernicus' corrections of the lunar elongation and anomaly appear to be a step in the wrong direction.

There are a few errors in Reinhold's computation of the lunar longitude. His value of c_3, 1;2,0°, should be, according to my calculations, 1;2,9°. While he does not state his values of c_4 and c_5, there must be an error, since it is not possible to obtain 0;7,35° for the lunar equation using 358;24,40° for the lunar anomaly. It is curious that Copernicus and Reinhold appear to make the same mistake at this point: Copernicus has 0;7° or 0;6,58° instead of 0;9,20°, and Reinhold has 0;7,35° rather than 0;11,13°. Reinhold finds the corrected elongation

$$\eta = \bar{\eta} + c_m = 87;12,13^o,$$

and using the mean longitude of the sun for the apparent, uncorrected time, his value of the true lunar longitude is

$$\lambda_m = \bar{\lambda}_s + \eta = 282;35,17^o.$$

There is no doubt that Reinhold found the declination of the moon, δ_m, as opposed to $\delta^*_m + \beta$. He gives a number of details from his computation. The question which arises, however, is the value assumed by Reinhold for the obliquity of the ecliptic at the time of the observation. It appears he originally assumed the minimum Copernican value, 23;28°, and computed a declination of 23;38,27°. Various changes made in the manuscript indicate that he then reomputed the declination assuming 23;29,1° for the obliquity, producing a declination of 27;38,28°. The value 23;29,1° is rather large for the obliquity at this time (1522). One would expect, from Copernican theory and Reinhold's parameters, 23;28,20°. In the second observation, Reinhold again appears to assume initially a value of 23;28° for the obliquity, and he then recomputes the declination with 23;29,0°.

Reinhold consciously compares his value of the declination with that of Copernicus, i.e., he writes his value of 27;38,28° and then says Copernicus had 27;41,0°. He uses 54;19,30° for the latitude of Frauenburg, rather than Copernicus' value of 54;19°. The lunar parallax is then (equations 1.1 and 1.3)

$$\pi = 82;50^o - (27;38,28^o + 54;19,30^o) = 0;52,2^o,$$

0;2,2° larger than the value obtained by Copernicus. However, if Copernicus had recomputed his declination in accordance with the corrections he made in the lunar longitude and latitude, his parallax would have been 0;53°. The value obtained from the parallax table in the Prutenic Tables is 0;50,24°, closer to Copernicus' value than to Reinhold's.

Table 1.6

REINHOLD'S RECOMPUTATION OF COPERNICUS' FIRST LUNAR PARALLAX OBSERVATION

Solar Longitude and the Equation of Time

apparent time since epoch	1521^j 269^d $17;40^h$
$\bar{p}$	26;46,30°
θ	166;18,47°
2θ	332;37,34°
c_p	0;32,49°
P	27;19,19°
λ_s^*	168; 3,45°
$\bar{k}$	96;24,23°
c_3	2; 0°
k	98;24,23°
c_4	1;49,57°
c_5	0;32,39°
c_6	0; 0,58°
$c_5 \cdot c_6$	0; 0,31,34°
c_s	1;50,29°
$\lambda_s^* - c_s$	166;13,16°
λ_s	193;32,35°
α_o	279;55,33°
α	192;28°
$\alpha - \alpha_o$	272;32,27°
$\bar{l}_s(t_o)$	278; 2,28°
$\bar{l}_s(t)$	194;50,15°
$\bar{l}_s(t) - \bar{l}_s(t_o)$	276;47,47°
E^o	−4;15,20°
E^h	−0;17^h
corrected time since epoch	1521^j 269^d $17;23^h$

Table 1.7

REINHOLD'S RECOMPUTATION OF COPERNICUS' FIRST LUNAR PARALLAX OBSERVATION

Lunar Longitude, Latitude, Declination and Parallax

	COPERNICUS Nuremberg edition	REINHOLD
z'	82;50°	82;50°
λ_s	193;29°	193;32,35°
$\bar{\eta}$	87; 6°	87; 4,38°
$2\bar{\eta}$		174; 9,16°
$\bar{\gamma}$	357;39°	357;22,40°
c_3		1; 2, 0°
γ	358;40°	358;24,49°
c_6		0;59,52°
c_m	0; 7°	0; 7,35°
η		87;12,13°
$\bar{\lambda}_s$		195;23, 4°
λ_m	282;33°	282;35,17°
$\bar{w}$	197; 1°	197; 2,32°
w	197; 8°	197;10, 7°
β	4;47°	4;46,36°
δ	27;41°	27;38,28°
ϕ_F	54;19°	54;19,30°
z	82°	81;57,58°
π	0;50°	0;52, 2°

3.2.3 Copernicus' Second Observation

The second observation reported by Copernicus was made in Frauenburg on August 7, 1525, six hours after noon. The time since epoch was 1524 Egyptian years, 234^d, 18^h. In the manuscript, Copernicus has twice changed the value of the observed zenith distance. He has originally written $81;43\frac{1}{2}^o$, changed this first to $81;42^o$, and then to 82°. The Nuremberg edition gives a fourth value for the apparent zenith distance, $81;55^o$.

The values given by Copernicus in the manuscript and the Nuremberg edition are listed in Table 1.8. The situation here is similar to that of the previous observation. After an initial computation of the true zenith distance and parallax, Copernicus has gone back and altered many of his values. In general, his changes are an improvement, i.e., his corrected values are closer to my own accurately recomputed values than his original values. The particular computations which led to the original manuscript values remain obscure.

Comparing the mean lunar motions in the two observations (Table 1.5), it appears that in both cases Copernicus used his second value for the argument of latitude ($\bar{\omega}_b$), which is not surprising since it is corrected in the text as well as in the tables. In both observations Copernicus used his original value of the lunar anomaly ($\bar{\gamma}_a$). Evidence for the elongation is inconclusive.

In the second observation, Copernicus' two values of the true lunar anomaly, along with his single value of the mean lunar anomaly, imply that the lunar correction c_3 is either

$$c_3 = 242;10^o - 239;43^o = 2;27^o$$

or

$$c_3 = 242;10^o - 239;38^o = 2;32^o.$$

The correct value, corresponding to $194;10^o$ for the double elongation, should be $2;30^o$. This last value agrees with the anomaly in the printed edition, $239;40^o$. All three values of the true anomaly will produce the same value for the lunar equation of center, $6;58^o$. While Copernicus states that his value for the lunar equation is 7°, his value of w reveals that he actually used $6;58^o$:

$$\begin{aligned} w &= \bar{w} + c_m \\ &= 193;19^o + 6;58^o = 200;17^o. \end{aligned}$$

He also uses $6;58^o$ for his second value of the true lunar longitude:

$$\begin{aligned} \lambda_m &= \bar{\lambda}_s + \bar{\eta} + c_m \\ &= 145;36^o + 97;5^o + 6;58^o = 249;39^o. \end{aligned}$$

This is assuming, however, that his value of the mean solar longitude was $145;36^o$, which is not certain. His first value for the true lunar longitude does not follow from the numbers he has stated.

Table 1.8

COPERNICUS' SECOND LUNAR PARALLAX OBSERVATION

	First version in Manuscript	Changes in Manuscript	Changes in Manuscript	Nuremberg edition
hours of time of observation	6^h			6^h
hours of apparent time since epoch	18^h			18^h
hours of corrected time since epoch	18; 4^h	18^h		18^h
z'	$81;43\frac{1}{2}^o$	$81;42^o$	82°	$\rightarrow 81;55^o$
λ_s	$144;21^o$	$144;14^o$		$144;14^o$
$\bar{\eta}$	$97;\ 5^o$			$\rightarrow 97;\ 6^o$
$\bar{\gamma}$	$242;10^o$			$242;10^o$
γ	$239;43^o$	$239;38^o$		$\rightarrow 239;40^o$
c_m	7°			7°
λ_m	$249;16^o$	$249;39^o$		$249;39^o$
$\bar{w}$	$193;19^o$			$193;19^o$
w	$200;17^o$			$200;17^o$
β	$4;41^o$			$4;41^o$
$\delta_m^* + \beta$	$26;23^o$	$26;36^o$		$26;36^o$
ϕ_F	$54;19^o$			$54;19^o$
z	$80;42^o$	$80;55^o$		$80;55^o$
π	$1;\ 1\frac{1}{2}^o$	$1;\ 5^o$		$\rightarrow 1^o$

The four possible alternatives for the lunar declination are as follows (see Figure 1.10b):

$$\begin{aligned}\delta_m^*(249;16^o)+4;41^o &= 26;32,24^o,\\ \delta_m^*(249;39^o)+4;41^o &= 26;35,51^o,\\ \delta_m(249;16^o,4;41^o) &= 26;30^o,\end{aligned}$$

and

$$\delta_m(249;39^o,4;41^o)=26;33^o.$$

Copernicus has written both 26;23° and 26;36° for the declination. It is not clear which was written first. There is no accounting for the 26;23°, but the value 26;36° seems a clear indication that Copernicus used δ_m^*, the declination of the point on the ecliptic which has the same longitude as the moon, with the addition of the lunar latitude. In the second observation, the lunar longitude is more than 20° from the winter solstice and the use of $\delta_m^* + \beta$ is not entirely satisfactory. The discrepancy between $\delta_m^* + \beta$ and δ_m introduces an error of 0;3° into the parallax. However, since Copernicus wavers in his observed zenith distance between 82° and 81; 42°, this hardly matters.

Copernicus' two values of the true zenith distance correspond to his two values of the declination (equation 1.2):

$$z=54;19^o+26;23^o=80;42^o$$

and

$$z=54;19^o+26;36^o=80;55^o.$$

The three values of the parallax can be explained as follows (equation 1.1):

$$\begin{aligned}\pi &= 81;43\frac{1}{2}^o-80;42^o=1;1\frac{1}{2}^o,\\ \pi &= 82^o-80;55^o=1;5^o,\end{aligned}$$

and

$$\pi=81;55^o-80;55^o=1^o.$$

Copernicus says that according to Ptolemy the parallax would have been 1;38°. Ptolemy's tables actually produce 1;34,38°. Copernicus' own parallax tables give 1;1,20° for π.

3.2.4 Reinhold's Recomputation of Copernicus' Second Observation

The values obtained by Reinhold in recomputing Copernicus' second observation are given in Tables 1.9 and 1.10. There seems to be a consistent pattern of error in Reinhold's values for both the solar and lunar correction c_3. In the first observation, I found, using the *Prutenic Tables*,

$$\begin{aligned}c_3(\theta) &= 1;59,35^o,\\ c_3(2\bar{\eta}) &= 1;2,9^o.\end{aligned}$$

Reinhold has 2;0° and 1;2,0°. In the second observation I found

$$c_3(\theta) = 1;57,55,$$
$$c_3(2\bar{\eta}) = 2;30,1^o.$$

Reinhold has 1;57,59° and 2;29,33°. There is definitely something peculiar about Reinhold's method for computing c_3.

Reinhold's manuscript reveals that he found a lunar declination δ_m of 26;33,59°, assuming the minimum obliquity of 23;28°. He then used an obliquity of 23;29,0° to obtain a declination of 26;34,54°. His value of the parallax is (equations 1.1 and 1.3)

$$\pi = 81;55^o - (54;19,30^o + 26;34,54^o) = 1;0,36^o.$$

The parallax tables in the *Prutenic Tables*, however, give 1;1,45°.

4. THE TRUE DISTANCE OF THE MOON IN TERRESTRIAL RADII

Copernicus and Reinhold find the true lunar distance from the parallax and apparent zenith distance:

$$S_m = \frac{\sin z'}{\sin \pi}.$$

Ptolemy, however, does not use the law of sines for oblique triangles. In order to solve triangle ADK (Figure 1.11) for the lunar distance KD, he divides the figure into two right triangles and solves for KL and LD. The details of Ptolemy's and Copernicus' numerical results, as well as Reinhold's recomputation of both, are listed in Tables 1.11 and 1.12.

4.1 Ptolemy's True Lunar Distance (Almagest *v.13)*

In Figure 1.11, AB is a great circle of the earth through the observer at A and GD is a circle through the moon D at the time of Ptolemy's observation. $EZHT$ is a circle to which the earth has the ratio of a point, similar in function to a great circle on the celestial sphere. The center of Ptolemy's circle, however, is the center of the earth, not the observer. The three circles lie in the same plane and have common center K. AZ is drawn parallel to KH, and AL, drawn perpendicular to DK, divides triangle ADK into two right triangles.

Ptolemy first solves triangle ADL for AL_1 and DL_1 with AD equal to 120.[19] He then finds AL_2 and KL_2 in triangle AKL for AK equal to 120. If AK, the radius of the earth, is then set equal to 1^{tr},

$$AL_3 = AL_2 \cdot \frac{1^{tr}}{120},$$

---

[19] The subscripts 1, 2, and 3 have been used to distinguish the values for 1) $AD = 120$, 2) $AK = 120$, and 3) $AK = 1^{tr}$.

Table 1.9

REINHOLD'S RECOMPUTATION OF COPERNICUS' SECOND LUNAR PARALLAX OBSERVATION

Solar Longitude and the Equation of Time

apparent time since epoch	$1523^j\ 219^d\ 18^h$
$\bar{P}$	26;48, 3°
θ	166;30,29°
2θ	333; 0,58°
c_p	0;32,23°
P	27;20,26°
λ_s^*	118;17,23°
$\bar{k}$	46;37,15°
c_3	1;57,59°
k	48;35,14°
c_4	1;21,15°
$c_5 \cdot c_6$	0; 0,22°
c_s	1;21,37°
$\lambda_s^* - c_s$	116;55,46°
λ_s	144;16,12°
α_o	279;55,33°
α	146;34,49°
$\alpha - \alpha_o$	226;39,16°
$\bar{l}_s(t_o)$	278; 2,28°
$\bar{l}_s(t)$	145; 5,26°
$\bar{l}_s(t) - \bar{l}_s(t_o)$	227; 2,58°
E^o	-0;23,42°
E^h	$-0;\ 1\frac{1}{2}^h$
corrected time since epoch	$1523^j\ 219^d\ 17;58,30^h$

Table 1.10

REINHOLD'S RECOMPUTATION OF COPERNICUS' SECOND LUNAR PARALLAX OBSERVATION

Lunar Longitude, Latitude, Declination and Parallax

	COPERNICUS Nuremberg edition	REINHOLD
z'	81;55°	81;55°
λ_s	144;14°	144;16,12°
$\bar{\eta}$	97; 6°	97; 5,11°
$2\bar{\eta}$		194;10,22°
$\bar{\gamma}$	242;10°	241;53,18°
c_3		2;29,33°
γ	239;40°	239;23,45°
c_6		0;59,15°
c_4		4;25,53°
$c_5 \cdot c_6$		2;34°
c_m	7°	6;59,53°
η		104; 5, 4°
$\bar{\lambda}_s$		145;37,49°
λ_m	249;39°	249;42,53°
$\bar{w}$	193;19°	193;19,33°
w	200;17°	200;19,26°
β	4;41°	4;41,15°
δ_m	26;36° †	26;34,54°
ϕ_F	54;19°	54;19,30°
z	80;55°	80;54,24°
π	1°	1; 0,36°

†For Copernicus this quantity is $\delta_m^* + \beta$.

Table 1.11

PTOLEMY'S TRUE LUNAR DISTANCE IN TERRESTRIAL RADII

		PTOLEMY	REINHOLD's Recomputation	REINHOLD's values converted to PTOLEMY's norm*
	$\angle ADL$	1; 7°	1;77°	
	$\angle DAL$		88;53°	
if	AD	120	10 000 000	120
then	AL	2;21	194 883	2;20,19
	DL	120	9 998 101	119;58,38
	$\angle AKL$	49;48°	49;48°	
	$\angle KAL$		40;12°	
if	AK	120	10 000 000	120
then	AL	91;39	7 637 960	91;39,20
	KL	77;27	6 454 577	77;27,22
	DL		391 850 985	
	KLD		398 305 562	
if	AK	1^{tr}	1^{tr}	
then	AL	$0;46^{tr}$		
	KL	$0;39^{tr}$		
	DL	$39;\ 6^{tr}$		
	KLD	$39;45^{tr}$	$39;49,50^{tr}$	

* These values do not appear in Reinhold's text.

Table 1.12

COPERNICUS' TRUE LUNAR DISTANCE IN TERRESTRIAL RADII

	First version in manuscript	Changes in manuscript	Changes in manuscript	Changes in manuscript	Nuremberg edition	Reinhold's recomputation
	First Observation (September 27, 1522)					
∠*DAE*	82;50°				82;50°	82;50°
∠*ACE*	82°				[82°]*	81;58°
∠*AEC*	0;50°				0;50°	0;52°
CE	99 027	99 219			99 219	9 921 874
AC	1 454				1 454	151 257
AC	1^{tr}				1^{tr}	1^{tr}
CE	68^{tr}				68^{tr}	$65;36^{tr}$
	Second Observation (August 7, 1525)					
∠*DAE*	81;43°	81;42°		82°	→81;55°	81;55°
∠*ACE*	80;42°	80;42°	$80;43\frac{1}{2}°$	80;55°	80;55°	[80;54,24°]
∠*AEC*	0;61°	0;60°		0;65°	→0;60°	1; 0,36°
CE	98 95[7]	98 953		99 027	→99 006	9 900 646
AC	1 7[74]	1 745		1 891	→1 747	176 259
AC	1^{tr}	1^{tr}		1^{tr}	1^{tr}	1^{tr}
CE	$56;47^{tr}$	$56;42^{tr}$		$[52;22^{tr}]$	$→56;41^{tr}$	$56;10^{tr}$

* Numbers in brackets are restored

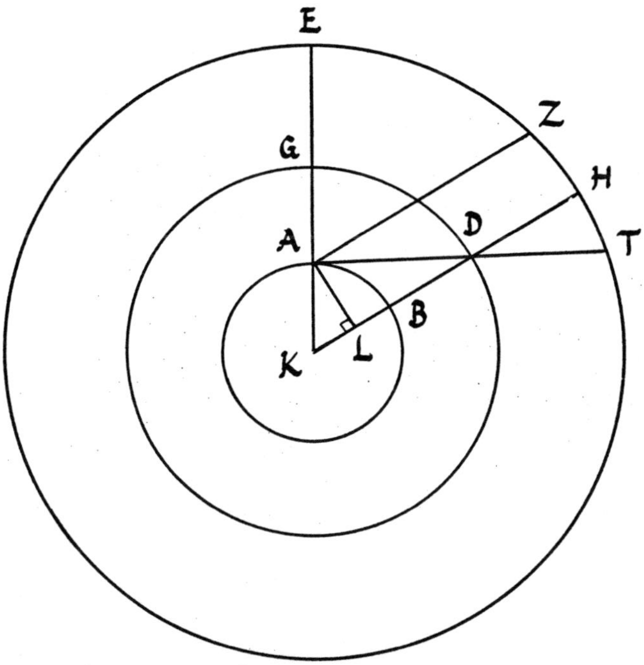

Figure 1.11: Ptolemy's True Lunar Distance in Terrestrial Radii

$$KL_3 = KL_2 \cdot \frac{1^{tr}}{120},$$

and

$$DL_3 = DL_1 \cdot \frac{AL_3}{AL_1}.$$

Finally, the lunar distance is the sum of KL_3 and DL_3.

Ptolemy has found for the lunar parallax

$$\pi = z' - z = 50;55^o - 49;48^o = 1;7^o.$$

He does not automatically equate the parallax with $\angle ADK$, however. Ptolemy states that the lunar parallax is measured by arc HT. Since the earth has the ratio of a point to circle $EZHT$, $\widehat{ZT}$ is not significantly greater than $\widehat{HT}$, and thus $\widehat{ZT}$ is approximately 1;7°. Also, the point A does not differ appreciably from the center K in relation to circle $EZHT$, and thus one may say

$$\angle ZAT = 1;7^o.$$

Then, since AZ and KH are parallel,

$$\angle ADK = \angle ZAT = 1;7^o.$$

Arc GD represents the true zenith distance, 49;48°.

In triangle ADL, with AD equal to 120,

$$AL_1 = \text{ crd } 2 \cdot 1;7^o = 2;20,19,12.$$

Ptolemy, however, has 2;21. Since the angle of parallax is small and the lunar distance is great, AD and DL differ by very little, and Ptolemy simply equates the two, i.e., he lets DL_1 equal 120. This is a reasonable approximation, since

$$DL_1 = \text{ crd } 2 \cdot (90^o - 1;7^o) = 119;58,37,58.$$

In triangle AKL, if AK is 120,

$$AL_2 = \text{ crd } 2 \cdot 49;48^o = 91;39,20$$

and

$$KL_2 = \text{ crd } 2 \cdot 40;12^o = 77;27,17,48.$$

Ptolemy has 91;39 for AL_2 and 77;27 for KL_2. If AK is then set equal to 1^{tr},

$$AL_3 = 91;39 \cdot \frac{1^{tr}}{120} = 0;45,49,30^{tr},$$

and

$$KL_3 = 77;27 \cdot \frac{1^{tr}}{120} = 0;38,43,30^{tr}.$$

Ptolemy has $0;46^{tr}$ for AL_3 and $0;39^{tr}$ for KL_3. Then

$$DL_3 = 120 \cdot \frac{0;46^{tr}}{2;21} = 39;8,56,10^{tr}.$$

Ptolemy has $39;6^{tr}$ for DL_3. Finally, the lunar distance is

$$DK = KL_3 + DL_3 = 0;39^{tr} + 39;6^{tr} = 39;45^{tr}.$$

If the computation is done more carefully, one obtains

$$\begin{aligned} DK &= KL + DL \\ &= \frac{\text{crd } 2(90^o - 49;48^o)}{2,0} + \frac{\text{crd } 2(90^o - 1;7^o)}{2,0} \cdot \frac{\frac{\text{crd } 2 \cdot 49;48^o}{2,0}}{\frac{\text{crd } 2 \cdot 1;7^o}{2,0}} \\ &= 0;38,43,38,54^{tr} + 39;12,2,8,52^{tr} = 39;50,45,47,46^{tr}. \end{aligned}$$

4.2 Reinhold's Recomputation of Ptolemy's True Lunar Distance

Reinhold repeats, step by step, Ptolemy's argument for equating $\angle ADL$ and the parallax of 1;7°. He modernizes the computation, however, by using sines instead of chords. Reinhold takes as a unit 10 000 000 in place of Ptolemy's 120. Then in triangle ADL, if AD is 10 000 000,

$$AL_1 = \sin 1;7^o = 194\ 883$$

and

$$DL_1 = \sin(90^o - 1;7^o) = 9\ 998\ 101.$$

In triangle AKL, if AK is 10 000 000

$$AL_2 = \sin 49;48^o = 7\ 637\ 960$$

and

$$KL_2 = \sin(90^o - 49;48^o) = 6\ 454\ 577.$$

At this point, Reinhold's method diverges slightly from Ptolemy's. Instead of finding AL, KL and DL when AK is 1^{tr}, Reinhold finds DL and KLD when AK is 10 000 000:

$$DL_2 = DL_1 \cdot \frac{AL_2}{AL_1} = 391\ 850\ 985$$

and

$$KLD_2 = KL_2 + DL_2 = 398\ 305\ 562.$$

This leaves only one conversion to be made for AK equal to 1^{tr}, namely

$$KLD_3 = KLD_2 \cdot \frac{1^{tr}}{10\ 000\ 000} = 39.8305562^{tr}.$$

Reinhold, of course, does not use decimals but writes this last step as

$$KLD_3 = 39;49,50^{tr}.$$

In the margin he has written Ptolemy's result. Table 1.11 lists Ptolemy's values in the first column, and the numbers found in Reinhold's manuscript in the second column. In order to facilitate comparison with Ptolemy, the third column contains Reinhold's values converted from a unit of 10 000 000 to one of 120. Following the completed calculation, Reinhold describes what he calls a simpler method. He draws a new diagram, eliminating circle $EZHT$ and perpendicular AL. It is the same as the diagram for Copernicus in Figure 1.12. With the radius of the earth equal to one, the lunar distance is simply

$$S_m = \frac{\sin 129;5^o}{\sin 1;7^o} = \frac{7\ 762\ 299}{194\ 883} = 39;49,50^{tr}.$$

*4.3 Copernicus' True Lunar Distance in Terrestrial Radii (*De Revolutionibus *iv.17)*

Copernicus finds the true lunar distance in terrestrial radii for both of his lunar zenith distance observations. His values are listed in Table 1.12. In Figure 1.12, AB is a great circle of the earth through the observer at A, and DE is a circle through the moon E at the time of the observation. The two circles lie in the same plane and have common center C. In the first observation, the lunar parallax, $\angle AEC$, is 0;50°, and the observed zenith distance of the moon, $\angle DAE$, is 82;50°. If AC is 1^{tr}, then the true lunar distance in terrestrial radii is

$$CE = \frac{\sin 82;50^o}{\sin 1;7^o} = \frac{99\ 219}{1\ 454} = 68;14,19^{tr}.$$

Copernicus has simply 68^{tr}. One can see from the manuscript that he first found sin82° using the true lunar zenith distance, rather than sin82;50° using the apparent zenith distance. A comparison of Tables 1.8 and 1.12 for the second observation reveals that Copernicus changed his values of the true and observed zenith distances in *De Revolutionibus* iv.17, as well as in iv.16. The changes are not entirely consistent, however. The values 81;43° for $\angle DAE$ and 1;1° for $\angle AEC$ are not found in chapter 16. The value 80;43$\frac{1}{2}$° for the true zenith distance (the third manuscript value in chapter 17), which, taken with the 81;43°, implies a parallax of 1;0$\frac{1}{2}$°, must be a combination of errors. In both chapters, the Nuremberg edition has a different set of values than the manuscript.

As in the first observation, the true lunar distance in terrestrial radii is

$$CE = \frac{\sin \angle DAE}{\sin \angle AEC}.$$

There are three different values for both sin$\angle DAE$ and sin$\angle AEC$ in the manuscript, plus a fourth value for each in the Nuremberg edition. They can be explained as follows (see Table 1.12):

$$\begin{aligned} 98\ 957 &= \sin 81;43^o, \\ 98\ 953 &= \sin 81;42^o, \\ 99\ 027 &= \sin 82^o, \\ 99\ 006 &= \sin 81;55^o, \\ 1\ 774 &= \sin 1;1^o, \\ 1\ 745 &= \sin 1;0^o, \\ \text{and} \qquad 1\ 891 &= \sin 1;5^o. \end{aligned}$$

Performing the divisions of the sine values to obtain the lunar distances, I find:

$$\frac{98\ 957}{1\ 774} = 56;46,54^{tr},$$

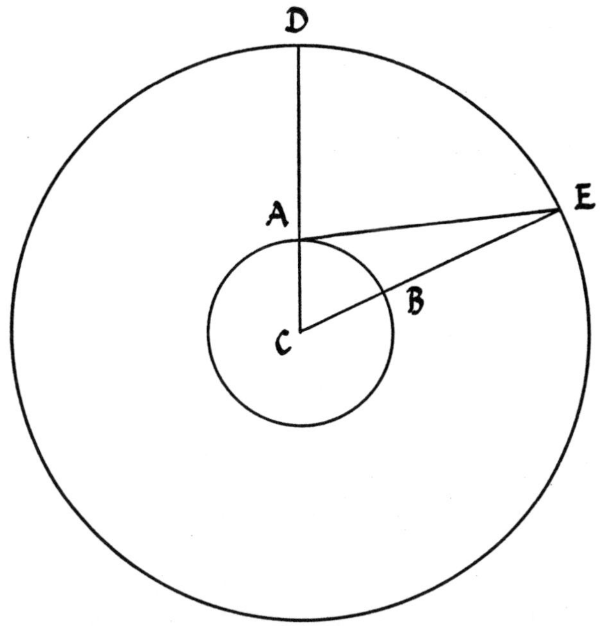

Figure 1.12: Copernicus' True Lunar Distance in Terrestrial Radii

$$\frac{98\ 953}{1\ 745} = 56;42,23^{tr},$$
$$\frac{99\ 027}{1\ 891} = 52;22,3^{tr},$$
$$\frac{99\ 006}{1\ 747} = 56;40,19^{tr},$$
$$\frac{99\ 006}{1\ 745} = 56;44,13^{tr},$$
and
$$\frac{99\ 027}{1\ 747} = 56;41,2^{tr}.$$

The first three results indicate that while Copernicus changed his values of angles DAE, ACE, AEC and their sines in the manuscript, he did not bother to recompute the division of the sine values, a computation which changes the lunar distance by more than 4^{tr}. The last three results indicate that the Nuremberg edition value of $56;41^{tr}$ for the true lunar distance does not follow either from Copernicus' stated values of the sines (99 006 and 1 747) or the correct values (99 006 and 1 745). These mistakes are particularly unfortunate in view of the fact that Copernicus subsequently discards the true lunar distance found in the first observation and uses only the results of the second observation.

4.4 Reinhold's Recomputation of Copernicus' True Lunar Distance

Using Copernicus' value of the observed zenith distance in the first ohservation, 82;50°, but his own recomputed value of the parallax, 0;52°, Reinhold finds the true lunar distance

$$CE = \frac{\sin 82;50^o}{\sin 0;52^o} = \frac{9\ 921\ 874}{151\ 257} = 65;35,46^{tr}$$

or 65;36^{tr}.[20]

Similarly, for the second observation, where Reinhold has found the parallax to be 1;0,36°,

$$CE = \frac{\sin 81;55^o}{\sin 1;0,36^o} = \frac{9\ 900\ 646}{176\ 269} = 56;10,4^{tr}$$

or 56;10^{tr}.

5. THE RELATIVE DISTANCE OF THE MOON

The next step is to find the relative distance of the moon from the lunar model. Since the original purpose of the lunar model is to generate the directions to the moon, and the tables have been set up to that end, a special calculation is needed to find the distance. Figure 1.13 shows the configuration of Ptolemy's model at the time of the observation. The observer is at E and the moon at L. The known parameters are the radius of the eccentric BD, the small circle radius DE, the epicycle radius BL, the double elongation $\angle DEB$, the equation of center $\angle BEL$, and the mean lunar anomaly $\widehat{PKL} = \angle PBL$. The values of these quantities suffice to determine the lunar distance at the time of the observation, assuming the mean distance at syzygy, AE, is 60.

Ptolemy proceeds by solving right triangles. DM and ZN have been drawn perpendicular to BE. In triangle DEM, $\angle DEM$ is $180^o - 2\bar{\eta}$, DE is given, and thus DM and EM are known. Since triangles EDM and EZN are congruent, NZ and EN are also known. In triangle BDM, BD and DM are given, and BM is determined by the Pythagorean theorem. Then

$$BE = BM - EM,$$

and

$$BN = BE - EN.$$

In triangle BNZ, BZ is found by the Pythagorean theorem, and $\angle NBZ$ can be found from the table of chords. Then in triangle BEL, which is not necessarily a right triangle, BE and BL are given, $\angle BEL$ is given and

$$\angle EBL = (\widehat{PKL} - 180^o) + \angle NBZ.$$

---

[20] Reinhold makes the same mistake as Copernicus in this computation in that he first writes the sine of $\angle ACE$, the true zenith distance.

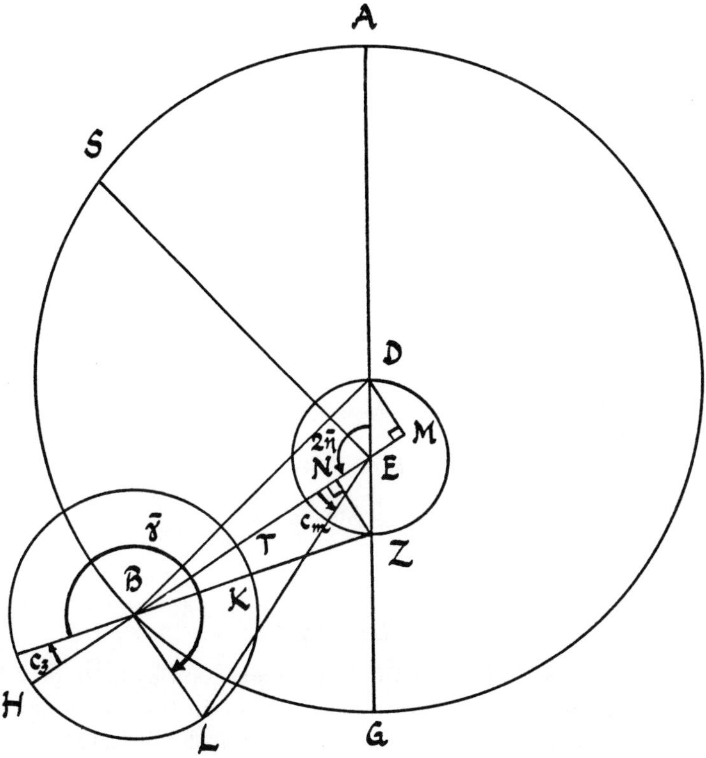

Figure 1.13: Ptolemy's Relative Lunar Distance

If $\angle EBL$ is not a right angle, a perpendicular must be dropped from B to EL and the two right triangles so formed can be solved for the component parts of EL, the lunar distance. For Ptolemy's observation, the value of $\angle EBL$ is very close to 90°, simplifying the solution. The details of Ptolemy's calculations are summarized below, and appear in the first column of Table 1.13.

Figure 1.14 shows the configuration of Copernicus' lunar model at the time of the second parallax observation. The observer is at E and the moon is at K. The known parameters are the deferent radius DE, the epicycle radius DC, the radius of the secondary epicycle CG, the double elongation $\widehat{FGK}$, the mean lunar anomaly $\widehat{ABC}$, the equation of center $\angle BEK$, and the lunar correction c_3, which is $\angle GDK$. In triangle DEK, $\angle DEK$ is given and

$$\angle EDK = \widehat{ABC} - \angle GDK - 180^\circ.$$

The relative lunar distance at the time of the observation, assuming a unit deferent radius (DE) is then simply

$$EK = 100\ 000 \cdot \frac{\sin \angle EDK}{\sin \angle DKE}.$$

The details of Copernicus' calculations, as well as Reinhold's recomputation, are listed in Table 1.14.

*5.1 Ptolemy's Relative Lunar Distance (*Almagest *v.13)*

The parameters of Ptolemy's lunar model are

$$DE = 10;19,$$

$$BD = 49;41,$$

and

$$BL = 5;15.$$

At the time of the observation

$$2\bar{\eta} = \angle DEB = 156;26^{o}\ ,$$

$$\bar{\gamma} = \widehat{PKL} = 262;20^{o}\ ,$$

and

$$c_m = \angle BEL = 7;40^{o}\ .$$

While Ptolemy must have computed the lunar correction c_3 ($\widehat{PH}$ or $\angle NBZ$) and the true lunar anomaly γ ($\widehat{HKL}$) in order to obtain the true lunar zenith distance, he does not state his values when discussing the observation. c_3 can be found from the lunar correction table as a function of $2\bar{\eta}$, but Ptolemy finds it here by solving triangle BNZ. In right triangles DEM and ZEN, DE and EZ are equal, and angles DEM and ZEN are equal. Thus the triangles are congruent, and

$$\angle DEM = \angle ZEN = 180^{o} - 156;26^{o} = 23;34^{o},$$

and

$$\angle EDM = \angle EZN = 90^{o} - 23;34^{o} = 66;26^{o}.$$

If

$$DE = EZ = 120,$$

then

$$DM = NZ = \text{ crd } 2 \cdot 23;34^{o} = 47;58,41$$

and

$$EM = EN = \text{ crd } 2 \cdot 66;26^{o} = 109;59,29.$$

Ptolemy has 47;59 for DM and 110;0 for EM. If

$$DE = EZ = 10;19$$

and

$$BD = 49;41$$

Table 1.13

PTOLEMY'S RELATIVE LUNAR DISTANCE

		PTOLEMY	REINHOLD's Recomputation	REINHOLD's values converted to PTOLEMY's norm*
	$\angle AES$	78;13°	78;13°	
	$\angle AEB$	156;26°	156;26°	
	$\angle DEM$	23;34°	23;34°	
	$\angle EDM$	66;26°	66;26°	
if	DE	120	10 000 000	120
then	DM	47;59	3 998 157	47;58,40
	EM	110; 0	9 165 955	109;59,29
if	DE	10;19	6 190	<u>10;19</u>
then	BD	49;41	29 810	<u>49;41</u>
	DM	4; 8	2 475	4; 7,30
	EM	9;27	5 674	9;27,24
	BD^2		888 636 100	
	DM^2		6 125 625	
	BM^2		882 510 475	
	BM	49;31	29 707	49;30,42
	MN		11 348	
	BE	40; 4		
	BN	30;37	18 359	30;35,54
	NZ	4; 8	2 475	
	BN^2		337 052 881	
	NZ^2		6 125 625	
	BZ^2		343 178 506	
	BZ	30;54	18 525	30;52,30

* Only those values which are underlined actually appear in Reinhold's text.

Table 1.13 (cont.)

		PTOLEMY	REINHOLD's Recomputation	REINHOLD's values converted to PTOLEMY's norm*
if	BZ	120	10 000000	120
then	NZ	16;2	1 336032	16; 2, 9
	$\angle NBZ$	7;40°	7;40,40°	
	$\widehat{PKL}$	262;20°	262;20°	
	$\widehat{KL}$	82;20°	82;20°	
	$\widehat{TKL}$	90°	90; 0,40°	
if	BL	5;15	3150	5;15
then	BE	40; 4	24033	40;3,18
	BE^2		577 585089	
	BL^2		9 922500	
	EL^2		587 577339	
	EL	40;25	24240	40;24

* These values do not appear in Reinhold's text.

Table 1.14

COPERNICUS' RELATIVE LUNAR DISTANCE

		COPERNICUS first version in manuscript	COPERNICUS Nuremberg edition	REINHOLD's Recomputation
	$\widehat{ABC}$	242;10°	242;10°	241;53,18°
	$\widehat{FGK}$	194;10°	→194;12°	194;10,22°
	$\angle GDK$	2;27°	→ 2;30°	2;29,33°
	$\angle KDB$	59;43°	→ 59;40°	59;23,45°
	$\angle CDB$	62;10°	62;10°	61;53,18°
	$\angle BEK$	7°	→ 12°	6;59,53°
	DE	91856	→91821	9 163 201
	EK	86354	→86310	8 607 050
if	DE	100000	100000	10 000 000
then	EK	94010	→93998	9 393 061

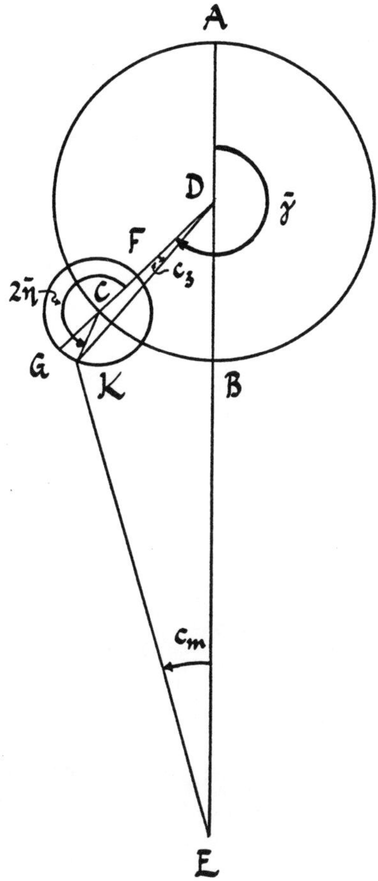

Figure 1.14: Copernicus' Relative Lunar Distance

then

$$DM = NZ = 47;59 \cdot \frac{10;19}{120} = 4;7,30,50,$$

and

$$EM = EN = 110;0 \cdot \frac{10;19}{120} = 9;27,25.$$

Ptolemy has 4;8 for DM and 9;27 for EM. In triangle BDM

$$BM = \sqrt{49;41^2 - 4;8^2} = 49;30,39.$$

Ptolemy may have used the approximation[21]

$$BM \cong 49;41 - \frac{4;8^2}{2 \cdot 49;41} = 49;30,41.$$

[21] Ptolemy usually states the values of the squared numbers when he uses the Pythagorean theorem, and he does not do that here. In what follows, the more accurately computed square root is given for comparison.

His stated value is 49;31. Then

$$BE = 49;31 - 9;27 = 40;4$$

and

$$BN = 40;4 - 9;27 = 30;37.$$

In triangle BNZ,

$$BZ = \sqrt{30;37^2 + 4;8^2} = 30;54,27,$$

or

$$BZ \cong 30;37 + \frac{4;8^2}{2 \cdot 30;37} = 30;53,45.$$

Ptolemy has 30;54. In the same triangle, if BZ is now set equal to 120,

$$\text{crd } 2 \cdot \angle NBZ = 4;8 \cdot \frac{2,0}{30;54} = 16;3,6.$$

Ptolemy, however, has 16;2. Using this value,

$$\angle NBZ = \frac{15;21,24^o}{2} = 7;40,42^o.$$

Ptolemy has 7;40°. Then

$$\widehat{KL} = 262;20^o - 180^o = 82;20^o$$

and

$$\widehat{TKL} = 82;20^o + 7;40^o = 90^o.$$

Since $\angle EBL$ is 90°, triangle BEL is a right triangle, and the relative lunar distance EL may be found by the Pythagorean theorem:

$$EL = \sqrt{40;4 + 5;15^2} = 40;24,32$$

or

$$EL \cong 40;4 + \frac{5;15^2}{2 \cdot 40;4} = 40;24,38.$$

Ptolemy has 40;25.

5.2 Reinhold's Recomputation of Ptolemy's Relative Lunar Distance

Reinhold's procedure for recomputing the relative lunar distance is virtually identical to that of Ptolemy. His results are listed in the second column of Table 1.13. The third column contains Reinhold's results converted to Ptolemy's norm. With the exception of the two underlined values of the lunar parameters DE and BD, the numbers in this column do not appear in Reinhold's manuscript. There is only one minor variation in procedure. After computing BM, Reinhold finds

$$MN = 2 \cdot EM$$

and

$$BN = BM - MN.$$

He does not compute BE until it is actually used in solving triangle BEL.

Reinhold finds the lunar correction c_3, $\angle NBZ$, to be 7;40,40°. In his recomputation of the zenith distance observation his stated value was only 7;40°. For Reinhold, $\angle EBL$ or $\widehat{TKL}$ is 90;0,40°, which he says is sufficiently close to 90° that triangle BEL may be treated as a right triangle.

There is only one error in Reinhold's work. The square of EL should be 587 507 589, and Reinhold has 587 577 339. The number is obtained simply by adding the squares of BE and BL. The difference does not affect the value of the relative lunar distance, 40;24.

*5.3 Copernicus' Relative Lunar Distance (*De Revolutionibus *iv.17)*

Copernicus finds the relative distance of the moon for his second observation only (see Figure 1.14). According to Copernicus' manuscript, at the time of the observation (see Table 1.14)

$$2\bar{\eta} = \widehat{FGK} = 194;10^\circ,$$

$$\bar{\gamma} = \widehat{ABC} = 242;10^\circ,$$

$$c_3 = \angle GDK = 2;27^\circ,$$ [22]

and

$$c_m = \angle BEK = 7^\circ.$$ [23]

In triangle DEK

$$\angle EDK = 242;10^\circ - 2;27^\circ - 180^\circ = 59;43^\circ$$

and

$$\angle DEK = 180^\circ - 7^\circ - 59;43^\circ = 113;17^\circ.$$

Then if the deferent radius DE is 100 000, the relative lunar distance is

$$\begin{aligned} EK &= \frac{\sin 59;43^\circ}{\sin 113;17^\circ} \cdot 100\,000 \\ &= \frac{86\,354}{91\,856} \cdot 100\,000 = 94\,010. \end{aligned}$$

[22] In the manuscript of ***De Revolutionibus*** iv.16, Copernicus has corrected the value of c_3 from 2;27° to 2;32. While other changes made in chapter 16 have been incorporated into chapter 17, this is one which Copernicus overlooked.

[23] While 7° is Copernicus' stated value of c_m, found in computing the zenith distance observation, he actually used the more accurate value of 6;58° in his computations in chapter 16.

The Nuremberg edition has a different set of values. There $\bar{\gamma}$ is still 242;10°, but

$$2\bar{\eta} = \widehat{FGK} = 194;12^o,$$

$$c_3 = \angle GDK = 2;30^o,$$

and

$$c_m = \angle BDK = 12^o.^{24}$$

Consequently, in triangle DEK,

$$\angle EDK = 242;10^o - 2;30^o - 180^o = 59;40^o$$

and

$$\angle DEK = 180^o - 12^o - 59;40^o = 108;20^o.$$

The relative lunar distance for unit deferent radius should then be:

$$EK = \frac{\sin\ 59;40^o}{\sin 108;20^o} \cdot 100\ 000 = \frac{86\ 310}{94\ 924} \cdot 100\ 000.$$

The Nuremberg edition does not have 94 924 for the sine of 108;20°, but 91 821. This indicates that the computation was not made with 12° for $\angle BEK$, but with the original 7°. For, if

$$\angle DEK = 180^o - 7^o - 59;40^o = 113;20^o,$$

then

$$\sin 113;20^o = 91\ 821.6.$$

The division of 86 310 by 91 821 does in fact yield a relative lunar distance of 93 998, Copernicus' value.

5.4 Reinhold's Recomputation of Copernicus' Relative Lunar Distance

In Reinhold's recomputation of Copernicus' second zenith distance observation he found

$$2\bar{\eta} = \widehat{FGK} = 194;10,22^o,$$

$$\bar{\gamma} = \widehat{ABC} = 241;53,18^o,$$

$$c_3 = \angle GDK = 2;29,33^o,$$

and

$$c_m = \angle BEK = 6;59,53^o.$$

In triangle DEK,

$$\angle EDK = 241;53,18^o - 2;29,33^o - 180^o = 59;23,45^o,$$

[24] This may be a misprint of xii for vii.

and

$$\angle DEK = 180^o - 6;59,53^o - 59;23,45^o = 113;36,22^o.$$

With a deferent radius DE of 10 000 000, Reinhold finds a relative lunar distance of

$$\begin{aligned} EK &= \frac{\sin\ 59;23,45^o}{\sin 113;36,22^o} \cdot 10\ 000\ 000 \\ &= \frac{8\ 607\ 050}{9\ 163\ 201} \cdot 10\ 000\ 000 = 9\ 393\ 061. \end{aligned}$$

Reinhold's values are listed in the last column of Table 1.14.

6. THE GREATEST AND LEAST DISTANCES OF THE MOON IN TERRESTRIAL RADII

For Ptolemy, if the deferent radius is AE, the epicycle radius BL, and the small circle radius DE, the mean and extreme distances are as follows:

distance in syzygy	
greatest	$AE + BL$
mean	AE
least	$AE - BL$
mean distance	$AE - DE$
distance in quadrature	
greatest	$AE - 2DE + BL$
mean	$AE - 2DE$
least	$AE - 2DE - BL$

For Copernicus, if the deferent radius is DE, the epicycle radius CD, and the secondary epicycle radius CF, the mean and extreme distances are:

distance in syzygy	
greatest	$DE + CD - CF$
mean	DE
least	$DE - (CD - CF)$
mean distance	DE
distance in quadrature	
greatest	$DE + CD + CF$
mean	DE
least	$DE - (CD + CF)$

Copernicus' lunar model was designed to reduce the variation in Ptolemy's lunar distance. This variation may be measured by finding the ratio of greatest to least distance S/s. In Ptolemy's model, the value of this ratio is close to 2. If the moon were in fact twice as far from the earth at its greatest distance as at its nearest approach, the lunar diameter would vary by a factor of two. Copernicus' model maintains the same mean distance at syzygy as at quadrature (see Tables 1.17 and 1.18), and reduces the ratio of greatest and least distances to 1;18,25. The greatest and least lunar distances both occur at quadrature where variation in diameter, area, and brightness is less noticeable. In Ptolemy's model, least distance occurs at quadrature, but the greatest distance occurs at conjunction, when the moon cannot be seen.

6.1 Ptolemy's Greatest and Least Lunar Distances (Almagest v.13)

Since Ptolemy found a true lunar distance of 39;45^{tr} and a relative distance of 40;25 for the time of his observation, multiplication by 39;45/40;25 ≅ 0;59 will convert the lunar parameters from a unit of 60 to terrestrial radii (see Table 1.15). Ptolemy converts only three parameters: the epicyclic radius

$$BL = 5;15 \cdot 0;59 = 5;10^{tr},$$

the mean distance at syzygy

$$AE = 60 \cdot 0;59 = 59^{tr},$$

and the mean distance at quadrature

$$EG = 39;22 \cdot 0;59 = 38;43^{tr}.$$

Ptolemy does not state the values of the greatest and least lunar distances in terrestrial radii until *Almagest* v.17, though he does mention and use the greatest distance in syzygy, 64;10^{tr}, in chapter 15. The ratio of greatest to least distance is 1;54,45. Ptolemy's greatest, least and mean distances in syzygy and quadrature are listed in the first column of Tables 1.17 and 1.18.

6.2 Reinhold's Recomputation of Ptolemy's Greatest and Least Lunar Distances

Reinhold found a relative lunar distance of 40;24 and a true distance of 39;45,50^{tr} for Ptolemy's observation. Thus, to convert from relative to true distances, he multiplies by

$$\frac{39;45,50}{40;24} = 0;59,3,19,$$

as opposed to Ptolemy's ratio of 0;59,0,37. Following Ptolemy, Reinhold converts only the three parameters: BL, AE and EG. His results are given in Table 1.15. Reinhold's value for the greatest distance in syzygy is 64;19,49^{tr}, as opposed to Ptolemy's 64;10^{tr}.

Reinhold and Ptolemy differ on the value of one parameter, EG, the mean distance in quadrature. Reinhold's value of 39;24, versus Ptolemy's 39;22, is stated more than once in the manuscript (as 23 640), and is the value used

Table 1.15

PTOLEMY'S CONVERSION FROM RELATIVE TO TRUE LUNAR DISTANCE

	PTOLEMY	REINHOLD's Recomputation	REINHOLD's values converted to PTOLEMY's norm*
Parameters of the Lunar Model			
BL	5;15	3 150	5;15
AE	60	36 000	60
EG	39;22	23 640	39;24
Relative Lunar Distance			
EL	40;25	24 240	40;24
True Lunar Distance in Terrestrial Radii			
EL	39;45^{tr}	143 390	39;49,50^{tr}
Lunar Parameters in Terrestrial Radii			
AE	59^{tr}	212 955	59; 9,15^{tr}
EG	38;43^{tr}	139 841	38;50,41^{tr}
BL	5;10^{tr}	18 634	5;10,34^{tr}
Greatest Lunar Distance			
$AE + BL$	64;10^{tr}	231 589	64;19,49^{tr}

* These values do not appear in Reinhold's text.

in computing EG in terrestrial radii. It is inconsistent with Reinhold's other values, however, since he has 60 for AE and 10;19 for DE, which imply

$$EG = AE - 2DE = 39;22.$$

This is an unusual error for Reinhold. It produces a slightly smaller value for the ratio of greatest to least lunar distance. If Reinhold's values of the lunar parameters had been consistent, there might have been reason to speculate on his having recomputed Ptolemy's parameters, perhaps in a section of the commentary now lost. This, however, seems unlikely.

Reinhold makes one reference to Ptolemy's values of the lunar distances. In the margin near his value of 59;9,15^{tr} for the mean distance in syzygy, he writes that Ptolemy has 59;0. There is a descriptive paragraph concluding the commentary on chapter 13 of the *Almagest*, but it says only that this is how Ptolemy found the lunar distances, and before finding the solar distance, one finds the angular diameters of the sun, moon and shadow.

*6.3 Copernicus' Greatest and Least Lunar Distances (*De Revolutionibus *iv.17)*

Copernicus, like Ptolemy, converts three lunar parameters into terrestrial radii: the deferent radius DE, and the sum and difference of the primary and secondary epicycle radii,

$$DFG = CD + CF$$

and

$$DF = CD - CF$$

(see Figure 1.14 and Table 1.16). In the manuscript, Copernicus' values for the lunar distance EK are 56;42^{tr} and 94 010. He converts

$$DE = 100\ 000,$$

$$DFG = 13\ 340,$$

and

$$DF = 8\ 600$$

into

$$DE = 60; 18^{tr},$$

$$DFG = 8; 2^{tr},$$

and

$$DF = 5; 11^{tr}.$$

The same results are given in the Nuremberg edition, where the values of EK are 56;41^{tr} and 93 998.[25]

[25] Using the manuscript values of EK and computing accurately, the values of DE, DFG and DF would be 60;19^{tr}, 8;3^{tr} and 5;11^{tr}, respectively. Using the values of EK in the printed edition, one obtains 60;18^{tr}, 8;3^{tr} and 5;11^{tr}. In computing the greatest and least distances, Copernicus in fact uses both 60;18^{tr} and 60;19^{tr} for DE, both in the manuscript and in the printed edition.

Table 1.16

COPERNICUS' CONVERSION FROM RELATIVE TO TRUE LUNAR DISTANCE

		COPERNICUS First version in manuscript	COPERNICUS Changes in Manuscript	COPERNICUS Nuremberg edition	REINHOLD's Recomputation
Parameters of the Lunar Model	DE	100 000		100 000	10 000 000
	DF	8 600		8 600	860 902
	DFG	13 340		13 340	
	CF				236 597
Relative Lunar Distance	EK	94 010		→93 998	9 393 061
True Lunar Distance in Terrestrial Radii	EK	56;42^{tr}		→56;41^{tr}	56;10^{tr}
Lunar Parameters in Terrestrial Radii	DE	60;18^{tr}		60;18^{tr}	59;48^{tr}
	DF	5;11^{tr}		5;11^{tr}	5; 9^{tr}
	DFG	8; 3^{tr}	8; 2^{tr}	8; 2	
	CF				1;25^{tr}
	FCG				2;50^{tr}
Greatest and Least Lunar Distances	$EDG = ED + DFG$	68;20^{tr}		68;20^{tr}	67;47^{tr}
	$ED - DFG$	52;15^{tr}	52;17^{tr}	52;17^{tr}	51;49^{tr}
	$EDF = ED + DF$	65;20^{tr}	65;30^{tr}	65;30^{tr}	64;57^{tr}
	$ED - DF$	55;15^{tr}	55; 8^{tr}	55; 8^{tr}	54;39^{tr}

Copernicus computes the greatest and least distances in syzygy and quadrature. The first set of values in his manuscript is so inconsistent it is hard to reconstruct what Copernicus might have had in mind. The results are given only in quarters (0;15) and thirds (0;20), but not always to the nearest of these. A revised set of values in the manuscript, which also appears in the Nuremberg edition (see Table 1.16), makes more sense, although it uses both $60;18^{tr}$ and $60;19^{tr}$ for DE. The greatest distance in quadrature is

$$EDG = 60;18^{tr} + 8;2^{tr} = 68;20^{tr},$$

the least distance in quadrature is

$$DE - DFG = 60;19^{tr} - 8;2^{tr} = 52;17^{tr},$$

the greatest distance in syzygy is

$$EDF = 60;19^{tr} + 5;11^{tr} = 65;30^{tr},$$

and the least distance in syzygy is

$$DE - DF = 60;19^{tr} - 5;11^{tr} = 55;8^{tr}.$$

The last column of Table 1.18 lists the values Copernicus would have obtained if he had used his first lunar zenith distance observation to find the lunar distances, rather than his second.

6.4 Reinhold's Recomputation of Copernicus' Greatest and Least Lunar Distances

Reinhold works with the radius of the secondary epicycle CF, rather than the sum of the epicycle radii DFG. His values for the lunar distance EK are $56;10^{tr}$ and 9 393 061, and his values of DF and CF also differ slightly from those of Copernicus. He converts DE from 10 000 000 to $59;48^{tr}$, DF from 860 902 to $5;9^{tr}$, and CF from 236 597 to $1;25^{tr}$. The greatest and least distances in syzygy and quadrature are then

$$\begin{aligned} EDG &= 59;48^{tr} + 5;9^{tr} + 2;50^{tr} = 67;47^{tr}, \\ DE - DFG &= 59;48^{tr} + (5;9^{tr} + 2;50^{tr}) = 51;49^{tr}, \\ EDF &= 59;48^{tr} + 5;9^{tr} = 64;57^{tr}, \end{aligned}$$

and

$$DE - DF = 59;48^{tr} - 5;9^{tr} = 54;39^{tr}.$$

Table 1.17

RELATIVE LUNAR DISTANCES – PTOLEMY, COPERNICUS, REINHOLD*

	PTOLEMY	REINHOLD Recomputation	COPERNICUS	REINHOLD Recomputation	COPERNICUS	REINHOLD Recomputation
SYZYGY						
Greatest	65;15	65;15	65; 9,36	65; 9,55,29	108 600	10 860 902
Mean	60; 0	60; 0	60	60	100 000	10 000 000
Least	54;45	54;45	54;50,24	54;50, 4,31	91 400	9 139 098
MEAN	49;41	49;42	60	60	100 000	10 000 000
QUADRATURE						
Greatest	44;37	44;39	68; 0,14,24	68; 0,15,28	113 340	11 334 096
Mean	39;22	39;24	60	60	100 000	10 000 000
Least	34; 7	34; 9	51;59,45,36	51;59,43,32	86 660	8 665 904
S/s	1;54,45	1;54,38	1;18,28	1;18,29	1;18,28	1;18,29

* None of these values appears explicitly in the texts of Ptolemy, Copernicus, and Reinhold, with the exception of the mean values.

Table 1.18

TRUE LUNAR DISTANCES IN TERRESTRIAL RADII – PTOLEMY, COPERNICUS, REINHOLD*

	PTOLEMY	REINHOLD Recomputation of Ptolemy	COPERNICUS Second Observation	REINHOLD Recomputation of Copernicus' Second Observation	COPERNICUS First Observation
SYZYGY					
Greatest	64;10^{tr}	64;19,49^{tr}	65;30^{tr}	64;57^{tr}	67;22^{tr}
Mean	59; 0^{tr}	59; 9,15^{tr}	60;18^{tr}	59;48^{tr}	62; 2^{tr}
Least	53;50^{tr}	53;58,41^{tr}	55; 8^{tr}	54;39^{tr}	56;42^{tr}
MEAN	48;51^{tr}	48;59,58^{tr}	60;18^{tr}	59;48^{tr}	62; 2^{tr}
QUADRATURE					
Greatest	43;53^{tr}	44; 1,15^{tr}	68;20^{tr}	67;47^{tr}	70;19^{tr}
Mean	38;43^{tr}	38;50,41^{tr}	60;18^{tr}	59;48^{tr}	62; 2^{tr}
Least	33;33^{tr}	33;40, 7^{tr}	52;17^{tr}	51;49^{tr}	53;45^{tr}
	1;54,45	1;54,38	1;18,25	1;18,29	1;18,30

* Only those values which are underlined appear in the texts

CHAPTER TWO

THE APPARENT DIAMETERS OF THE SUN, MOON, AND SHADOW

Since the solar parallax is small and impossible to determine directly by naked eye observation, it was not feasible for ancient and medieval astronomers to calculate the solar distance from the parallax. Ptolemy and Copernicus use an alternate geometric method which depends on the angular diameters of the sun, the moon, and the shadow of the earth.

In Figure 2.1, the center of the sun is at D, the center of the moon is at L, and the center of the earth is at K. The true diameters of the sun, moon, and earth are AC, NO, and GE, respectively. The angular diameter of an object, also called the apparent diameter, is the angle subtended at the observer's eye by the diameter of the object. Angular diameter varies as a function of the distance of the object from the observer – the closer the object the larger its apparent diameter. The difference in apparent diameter as seen from the center of the earth and as seen by an observer on the surface of the earth is small and is ignored for the present purposes.[1] Thus, in Figure 2.1, the apparent diameter of the moon (R_m) is $\angle NKO$, even though the observer may be at F. The shadow cone cast by the earth will intersect a plane perpendicular to the axis (KS) of the cone, at the distance $KM = KL$, in a circle of diameter QR. This circle is what I shall call the shadow. The apparent radius of the shadow (R_{sh}) at distance KM is $\angle MKR$. The geometric method for determining the solar distance utilizes the near equality of the apparent diameters of the sun and moon. If the solar distance is fixed and if KL is the lunar distance at which solar and lunar diameters appear equal, then the solar distance in terrestrial radii can be derived from

1. the angular radius of the moon at distance KL
2. the angular radius of the shadow at this same distance, and
3. the value of the lunar distance in terrestrial radii.

This chapter concerns the calculation of the apparent radii of the moon and the shadow at the given lunar distance. This can be done with some precision by using two lunar eclipse observations which occur when the moon is approximately at the distance KL.

[1] If the true lunar radius is $0;17^{tr}$ (Reinhold's value), then the difference between the angular lunar diameters as seen from 60^{tr} and 59^{tr} is less than $0;0,30^{o}$.

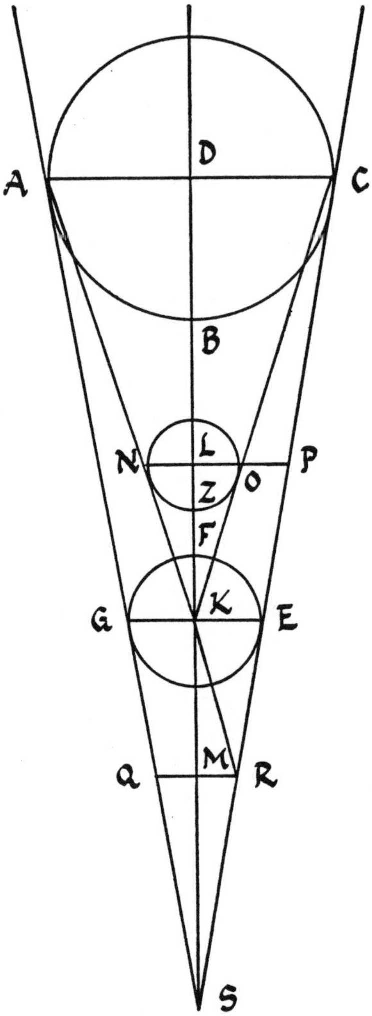

Figure 2.1: The Solar Distance

Figure 2.2 represents the configuration of the shadow (with center A) and the moon (with center B) at the time of mid-eclipse. The distance of the moon from the lunar node (BN) can be determined from the lunar model. Since the inclination of the lunar orbit (i) is known, one can find the distance AB between the centers of the moon and the shadow on a great circle perpendicular to the lunar orbit. This distance is equal to the sum of the angular radii of the moon and the shadow, minus that fraction of the moon's diameter which is eclipsed:

$$AB = R_m + R_{sh} - (m/12) \cdot 2R_m, \tag{2.1}$$

where m is the magnitude of the eclipse in digits. Since there are two unknowns

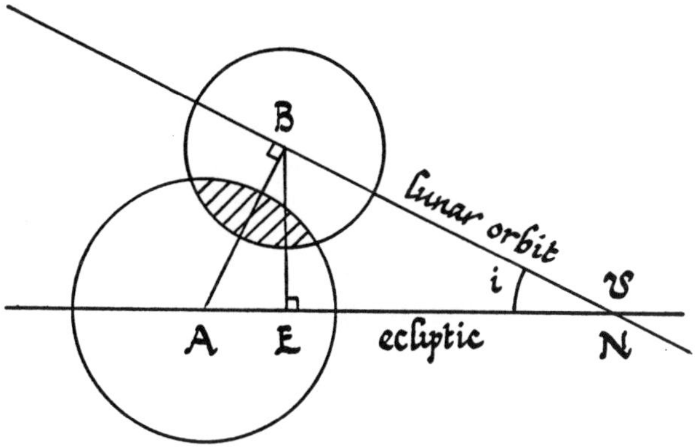

Figure 2.2a: The First Eclipse

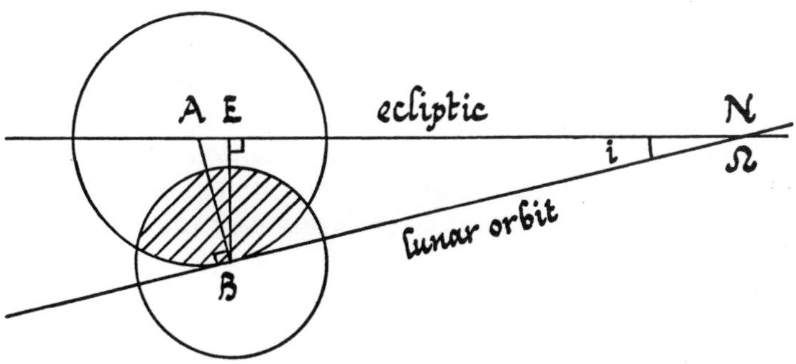

Figure 2.2b: The Second Eclipse

(R_m and R_{sh}), two eclipses of unequal magnitude will suffice to determine the radii of the moon and the shadow.

1. PTOLEMY (*Almagest* v.14)

In a discussion preceding his computations (*Almagest* v.14), Ptolemy mentions that he considers the use of water clocks and rising times inadequate for determining the angular diameters of the sun and moon.[2] He mentions the dioptra described by Hipparchus,[3] which he says he used only to determine 1) that the

[2] In his commentary on this portion of the *Almagest*, Reinhold reiterates Ptolemy's rejection of the use of water clocks and rising times. He cites several sources on the subject, including Theon, Cleomedes, Macrobius, and Martianus Capella. He also refers to his "Scholium in the *Theoricae* of Peurbach" (Ms.R.177r), where he mentions Cleomedes and Macrobius (Peurbach [1], f. 205v).

[3] On the dioptra of Hipparchus, see Rome [1], p.87ff., Proclus [1], pp.119-131, and Hultsch [1], pp.193-209.

solar diameter does not vary appreciably in the course of a year, and 2) that the solar and lunar diameters appear equal when the moon is at its greatest distance in syzygy ($64;10^{tr}$).[4] Ptolemy goes on to say that earlier astronomers found the solar and lunar diameters equal when the moon was at its mean distance and that his values are smaller than those found by his predecessors.[5]

In order to find the apparent lunar diameter when the solar and lunar diameters appear equal, Ptolemy must use two lunar eclipses in which the moon is near its greatest distance in syzygy. For this purpose, he chose two ancient eclipses, recorded by the Babylonians.

1.1 The First Eclipse Observation

Ptolemy states that the first lunar eclipse began at the end of the eleventh hour in Babylon in the fifth year of Nabopolassar, the 127th year of Nabonassar, the night of 27/28 Athyr, and that the moon was eclipsed one quarter of its diameter from the south. The date of this eclipse is April 22, –620. Ptolemy interprets "end of the eleventh hour" to mean 5 seasonal hours after midnight. He goes on to say that the middle of the eclipse was 6 seasonal hours after midnight, or 5 5/6 equinoctial hours in Babylon. This implies that Ptolemy assumed the half-length of night in Babylon at this time was 5;50 equinoctial hours, and that the half-duration of the eclipse was one seasonal hour or 0;58,20 equinoctial hours.

Using the solar mean motion and correction tables in the *Almagest*, I have found that the true solar longitude 6^h after midnight in Babylon on the day in question was approximately $27;3^o$. Assuming the latitude of Babylon is 35° and using the tables in *Almagest* ii.8, the half-length of night at this time of year is $5;30^h$, not $5;50^h$ as Ptolemy states.[6]

The duration of an eclipse can be estimated from its magnitude. Manitius suggests that on this basis Ptolemy's value of $0;58,20^h$ is reasonable.[7] Using Ptolemy's eclipse tables (*Almagest* vi.8), I have computed a half-duration time of $1;2^h$. I believe Ptolemy's value of $0;58,20^h$ actually represents a convenient

[4] This excludes the possibility of an annular solar eclipse.

[5] These remarks may be applied to Hipparchus. Cf. Swerdlow [2], pp. 287-305.

[6] There is a question here of what I should use for the latitude of Babylon. On the one hand, Ptolemy states in *Geography* v.19 (Ptolemy [3], p. 1021) that the lattitude of Babylon is 35°. The Babylonian value for the ratio of the longest day to the shortest night is 3:2, which corresponds to a longest day of 14 $2/5^h$. This agrees well with Ptolemy's values for the latitude of Babylon, since 35° corresponds to 14.41^h for the length of the longest day, where $30;22^o$ corresponds to 14^h and 36° to 14 $1/2^h$. In *Almagest* xiii.7, Ptolemy says that the observations of the Chaldeans were made near the parallel of Phoenicia, where the longest day is 14 $1/4^h$. Using Ptolemy's tables in *Almagest* ii.8, I find for a latitude of 35° that the half-length of daylight should be $5;30^h$. On the other hand, by a straight linear interpolation half-way between the values in Ptolemy's tables for 14^h and 14 $1/2^h$, i.e., for a latitude of approximately $33;11^o$, I find a half-length of daylight of $5;53^h$, very close to Ptolemy's value of $5;50^h$. I may add that the latitude of Babylon is in fact about 32 $1/2^o$.

[7] Ptolemy [1], v. 1, p. 307, note a.

combination of the conversion of seasonal to equinoctial hours and the estimation of the time of half-duration.[8]

Since the time of mid-eclipse in Babylon was 5;50^h after midnight, and Ptolemy assumes a time difference between Babylon and Alexandria of 0;50^h the time in Alexandria was 5^h after midnight. The magnitude of an eclipse at this time, computed from Ptolemy's table of lunar eclipses, is indeed 3 digits.[9] The actual time of mid-eclipse in Alexandria, re-determined from modern tables, was 4;36^h after midnight,[10] and the eclipse magnitude was 2.1 digits.[11] This implies that Ptolemy's time is in error by –0;24^h, and his eclipse magnitude is in error by –0.9 digits.

1.2 The Second Eclipse Observation

The second observation is of special interest because it is the only eclipse mentioned by Ptolemy for which a corresponding Babylonian report survives in a cuneiform text.[12] Ptolemy states that in the year 7 of Cambyses, 17/18 Phamenoth, the moon was eclipsed from the north by one half of its diameter one hour before midnight in Babylon. The date of this eclipse is July 16, –522. The description of the eclipse in the Bablyonian text is as follows:

> Year VII, month IV, night of the 14th, 1 2/3 double hours in the night a 'total' lunar eclipse took place [with only] a little remaining [uneclipsed]. The north wind blew.[13]

Ptolemy goes on to say that the eclipse took place 1;50 equinoctlal hours before midnight in Alexandria and he uses this value for the time of mid-eclipse in his calculations. It appears that Ptolemy by "one hour before midnight in Babylon" means that the time of mid-eclipse is one equinoctial hour before midnight.

The information about this eclipse which was available to Ptolemy may or may not have included a time of 1 2/3 double hours. Britton has determined[14] that "1 2/3 double hours after the beginning of the night" corresponds to 22;34^h after noon, whereas the apparent time in Babylon at the beginning of

---

[8] According to Britton [1], p. 81, the time of half-duration is 0;51^h. Oppolzer [1], p. 334, gives 0;42^h.

[9] With a lunar anomaly of 340;7° and true argument of latitude 80;40°, $m = 2;55+(5;4-2;56)\cdot 0;2,1 = 3;0,18$.

[10] Britton [1], p. 81.

[11] Britton [1], p. 81. Oppolzer [1], p. 334, has 1.6 digits.

[12] The text is Strm. kambys. 400 rev., published by Kugler [1], p. 71.

[13] This translation of Kugler's transcription was provided by Professor Sachs in a private communication to John Britton (cf. Britton [1], p. 82). Kugler's incorrect translation is of interest, however, because of its agreement with Ptolemy: Year 7, month IV, night of the 14th, 1 2/3 double hours after nightfall a lunar eclipse; the whole course is visible; it extended itself over the northern half of the [moon's] disk.

[14] Britton [1], p. 83.

the eclipse was 22;35^h. As he remarks, however, the close agreement may be strictly fortuitous. The modern value for the apparent time of eclipse middle in Alexandria is 22;58^h.[15] Ptolemy's value of 22;10^h is thus in error by +0;48^h. The actual magnitude of the eclipse was 6.1 digits.[16] It appears then that while Ptolemy's description of the eclipse magnitude is very good, his time is greatly in error, whereas the time in the Babylonian text may be quite exact, but the magnitude is clearly wrong.

1.3 The Computation of the Positions of the Moon in the Two Eclipses

The calculation of the solar and lunar longitudes and the lunar latitudes are described in detail in Appendix B. Tables 2.1 and 2.2 list the results obtained by Ptolemy and Reinhold in his commentary.

In the Ptolemaic model, greatest distance in syzygy occurs when the moon is at the apogee of its epicycle in syzygy. In the first eclipse the moon was 19;53° from the epicycle's apogee, and in the second 28;5°. I assume Ptolemy chose to use two Babylonian eclipses because in his own time he did not observe two eclipses which satisfied his conditions.[17]

1.4 The Calculation of the Diameters

Figure 2.2 illustrates the configuration of the moon and the earth's shadow in the two eclipses. AN is the ecliptic, BN is the lunar orbit, A is the center of the earth's shadow, B is the center of the moon, N is the lunar node, descending in the first eclipse and ascending in the second. Figure 2.2a shows the moon eclipsed from the south by one quarter of its diameter, as in the first observation, and Figure 2.2b shows the moon eclipsed from the north to the extent of one half its diameter, as in the second. BE, perpendicular to the ecliptic, measures the latitude of the moon. Ptolemy states very explicitly that he wishes to find the distance AB between the centers of the moon and the earth's shadow on the great circle through the poles of the lunar orbit. In order to compute this quantity, one needs the distance from the center of the moon to the lunar node, BN. In the first eclipse, Ptolemy found that the true argument of latitude was 80;40°, and thus

$$BN = w_{☋} - w = 90^o - 80;40^o = 9;20^o.$$

In the second eclipse, where Ptolemy found 262;12° for the true argument of latitude,

$$BN = w_{☋} - w = 270^o - 262;12^o = 7;48^o.$$

[15] Britton [1], p. 83.

[16] Britton [1], p. 83, and Oppolzer [1], p. 335.

[17] Between A.D. 125 and 141, I have found no more than thirteen lunar eclipses which may have been visible in Alexandria. The closest approaches of the moon to the epicycle's apogee in these eclipses were between 50° and 60°.

Table 2.1

THE FIRST ECLIPSE OBSERVATION

	PTOLEMY	REINHOLD's Recomputation
apparent time since epoch	$126^a\ 86^d\ 17^h$	$126^a\ 86^d\ 17^h$
$\bar{k}$		320; 4,45°
c_s		1;28,20°
k		321;33, 5°
λ_s	27; 3°	27; 3, 5°
$\bar{\lambda}_s$		25;34,45°
$\bar{\lambda}_s(t_o)$		330;45°
$\bar{\lambda}_s(t) - \bar{\lambda}_s(t_o)$		54;49,45°
$\alpha(27;3,5)$		25; 3,30°
α_o		335; 7°
$\alpha - \alpha_o$		49;56,30°
E^o		4;53,15
E^h	[$-0;15^h$]*	$-$ 0;19,30^h
corrected time since epoch	$126^a\ 86^d\ 16;45^h$	$126^a\ 86^d\ 16;45^h$
$\bar{\lambda}_m$	205;32°	205;32,58°
$\bar{\eta}$		179;58,13°
$2\bar{\eta}$		359;56,26°
$\bar{\gamma}$		340; 5, 6°
$\bar{\text{w}}$	[79; 7°]	79; 3,34°
γ	340; 7°	
c_m	[1;33°]	1;34°
λ_m	207; 5°	207; 6,58°
w	80;40°	80;37,34°
AB†	0;48,30°	0;48,47°

* Numbers in brackets do not appear in the text.

† For Reinhold, this quantity is β, not AB.

Table 2.2

THE SECOND ECLIPSE OBSERVATION

	PTOLEMY	REINHOLD's Recomputation
apparent time since epoch	$224^a\ 196^d\ 10;10^h$	$224^a\ 196^d\ 10;10^h$
$\bar{k}$		44;22,44°
c_s		1;37°
k		42;45,44°
λ_s	108;12°	108;15,44°
$\bar{\lambda}_s$		109;52,44°
$\bar{\lambda}_s(t_o)$		330;45°
$\bar{\lambda}_s(t) - \bar{\lambda}_s(t_o)$		139; 7,44°
$\alpha(108;15,44)$		109;50°
α_o		335; 7°
$\alpha - \alpha_o$		134;43°
E^o		4;24,44°
E^h	[$-0;20^h$]*	$-0;17,30^h$
corrected time since epoch	$224^a\ 196^d\ 9;50^h$	$224^a\ 196^d\ 9;50^h$
$\bar{\lambda}_m$	290;22°	290;20,46°
$\bar{\eta}$		180;28, 2°
$2\bar{\eta}$		0;56, 4°
$\bar{\gamma}$		27;54,18°
$\bar{\mathrm{w}}$	[264;20°]	264;21,38°
c_3		0; 9°
γ	28; 5°	28; 3,18°
c_m	[2; 8°]	2;10,20
λ_m	288;14°	288;10,26°
w	262;12°	262;11,18°
AB†	0;40,40°	0;40,41°

* Numbers in brackets do not appear in the text.
† For Reinhold, this quantity is β, not AB.

Ptolemy states that in the first eclipse

$$AB = 0;48,30^o$$

and in the second

$$AB = 0;40,40^o,$$

with no indication of how he arrived at these results. I have not been able to reproduce Ptolemy's values.[18]

In Reinhold's recomputation of this passage, his values for the true argument of latitude, 80;37,34° in the first eclipse and 262;11,18° in the second, differ slightly from Ptolemy's and thus produce different nodal distances:

$$BN = 90^o - 80;37,34^o = 9;22,26^o$$

in the first eclipse, and

$$BN = 270^o - 262;11,18^o = 7;48,42^o$$

in the second. Reinhold states that he finds the lunar latitude, i.e., BE, obtaining in the first eclipse

$$BE = 0;48,47^o$$

and in the second

$$BE = 0;40,41^o.$$

I have solved spherical triangle BEN and found 0;48,48° and 0;40,43° for the latitudes in the two eclipses. These are also the values I obtained from Reinhold's latitude table in the *Prutenic Tables.* In his subsequent discussion of the diameters of the moon and shadow, Reinhold describes this particular distance as perpendicular to the lunar orbit, but he also calls it the latitude. It appears that he considered the latitude, measured perpendicular to the ecliptic, a sufficiently good estimate of AB, the distance between the centers of the moon and the shadow, measured perpendicular to the lunar orbit. In fact, I believe Reinhold thought Ptolemy himself found only the latitude. This cannot be confirmed, however, since Reinhold does not use his own values for the latitude in the calculation of the lunar and shadow diameters. After his elaborate recomputations, Reinhold merely uses Ptolemy's values of AB to reproduce Ptolemy's results for the diameters.

Substituting Ptolemy's values of AB into equation 2.1, Ptolemy and Reinhold both find

$$R_m + R_{sh} - 1/4 \cdot 2R_m = 0;48,30^o$$

in the first eclipse, and

$$R_m + R_{sh} - 1/2 \cdot 2R_m = 0;40,40^o \qquad (2.2)$$

[18] Using Menelaus' theorem, Pappus recomputed AB for the first eclipse and obtained 0;48,30°. Rome [1], pp. 97-103.

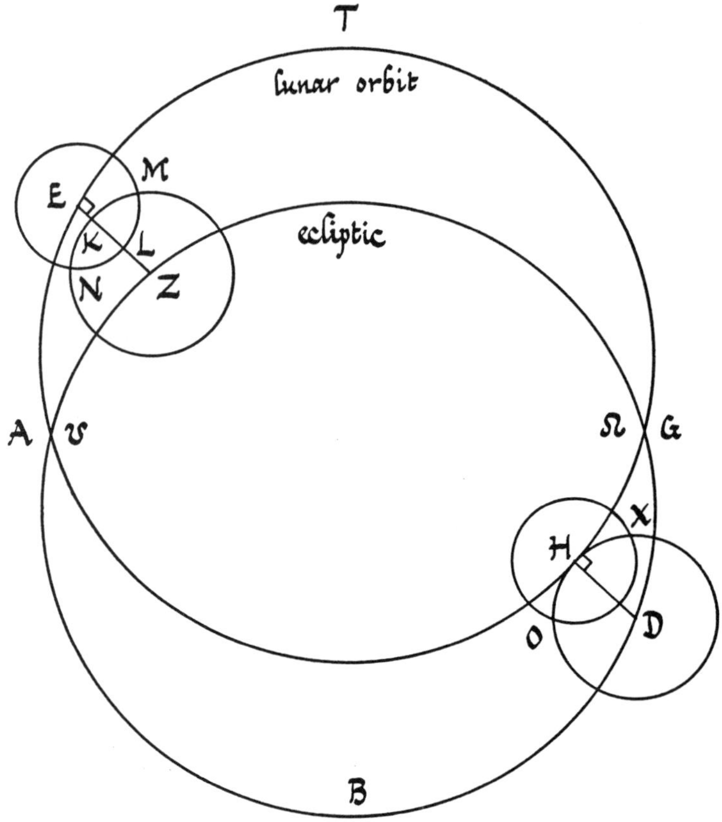

Figure 2.3: Reinhold's Illustration of the Two Eclipses

in the second. Subtracting the second equation from the first,

$$1/4 \cdot 2R_m = 0;7,50^o,$$

and thus the apparent lunar radius is 0;15,40° and the apparent lunar diameter is 0;31,20°. Since the magnitude of the second eclipse was one half, equation 2.2 gives the apparent shadow radius as 0;40,40°.

Figure 2.3 reproduces the diagram in Reinhold's manuscript (Ms.R.179r) which is used to illustrate his calculation of the lunar and shadow radii. In the first eclipse the moon is at E, and in the second eclipse the moon is at H. This figure is almost identical to the diagram used by Pappus in his commentary on *Almagest* v.14.[19]

In the determination of the solar distance, Ptolemy and Copernicus use the ratio of the lunar and shadow radii to derive the true shadow radius (MR)

[19] Rome [1], p. 97. The lettering used by Pappus and Reinhold is the same but the circles of the ecliptic and lunar orbit have been interchanged. The diagram used by Regiomontnaus in *Epitome* v.18 is quite different. In Figure 2.3, the Greek letters have been changed to corresponding Roman letters.

from the true lunar radius (LO):[20]

$$MR = LO \cdot \frac{R_{sh}}{R_m}.$$

For Ptolemy, the moon-shadow radius ratio, frequently referred to below as "the ratio," is

$$\frac{R_{sh}}{R_m} = \frac{0;40,40^o}{0;15,40^o} = \frac{2442''}{940''} = \frac{122}{47}\ .$$

Ptolemy uses the approximate value of 2 3/5, which he says is a little large. Reinhold compares 122:47 to Ptolemy's value of 13:5 by writing

$$\frac{122}{47} = \frac{13}{5\ 1/122}\ .$$

2. COPERNICUS (*De Revolutionibus* iv.18)

The only information which Copernicus gives concerning his two eclipses is that in the first, 3/12 of the diameter of the moon was eclipsed and the latitude was 0;47,54°, and in the second, 10/12 of the diameter was eclipsed and the latitude was 0;29,37°. Not only are there no details of the reduction of the observations, but even the dates of these eclipses are not given. I believe they are purely fictitious.

Reinhold's only comment on *De Revolutionibus* iv.18 is that it is sufficiently clear in itself.[21] While this is not true in light of the unknown source of the observations, it is typical for Reinhold to avoid matters which he could not explain. It is true, however, that given the information Copernicus presents, his computation of the radii of the moon and shadow is straight-forward and correct.

Copernicus does not refer to the distinction between AB and BE, and does not hesitate to use the lunar latitude. Substituting in equation 2.1, but with BE in place of AB,

$$R_m + R_{sh} - 3/12 \cdot 2R_m = 0;47,50^o \tag{2.3}$$

in the first eclipse, and

$$R_m + R_{sh} - 10/12 \cdot 2R_m = 0;29,37^o$$

in the second. Subtracting the second equation from the first,

$$7/12 \cdot 2R_m = 0;18,17^o.$$

---

[20] Strictly speaking,

$$MR = LO \cdot \frac{\tan R_{sh}}{\tan R_m}\ ,$$

where $KL = KM$.

[21] "Satis ex sese perspicuum," (Ms.R.53r).

The apparent lunar radius is thus 0;15,40° and the diameter is 0;31,20°. Equation 2.3 is equivalent to

$$R_{sh} = 0;47,54^o - 1/2R_m$$

and thus

$$R_{sh} = 0;47,54^o - 0;7,50^o = 0;40,4^o.$$

Copernicus also remarks that

$$\begin{aligned} R_{sh} &= 0;29,37^o + 1/3D_m \\ &= 0;29,37^o + 0;10,27^o = 0;40,4^o, \end{aligned}$$

as if this were additional confirmation of his results, which it is not. The diameter of the shadow is then 1;20,8°. Ptolemy had found 0;40,40° and 1;21,20° for the radius and diameter of the shadow.

If Copernicus' observations are indeed of his own invention, then, by the proper choice of magnitude and latitude, he could produce any desired diameter. His lunar diameter is in fact identical to Ptolemy's. Copernicus remarks on this and specifies that his values for the lunar and shadow diameters, 0;31,20° and 1;20,8°, are for the greatest lunar distance in syzygy. In the following chapter *(De Revolutionibus* iv.19), he gives a different set of radii for this distance, namely 0;30° and 1;20,40°. This second set of values can be partially explained as an adjustment of Ptolemy's radii to the new lunar distances of Copernicus. The choice of values in chapter 18, however, remains somewhat of a mystery.

CHAPTER THREE

THE DISTANCE OF THE SUN FROM THE EARTH

In the Ptolemaic solar model, the distance of the sun from the earth varies between a greatest distance in apogee and a least distance in perigee. In *Almagest* v.14. Ptolemy states that he did not find the diameter of the sun to vary with its distance. When he finds a solar distance in *Almagest* v.15 of 1210^{tr}, he does not specify whether this is the greatest, least or mean distance. From the *Planetary Hypotheses* we know that Ptolemy considered 1210^{tr} as the mean solar distance, and that the greatest and least solar distances were 1260^{tr} and 1160^{tr}.[1] In the *Narratio Prima*, Rheticus states that Ptolemy's value of 1210^{tr} is for the greatest solar distance.[2] Copernicus and Reinhold appear to make the same assumption.

Copernicus assumes that the apparent solar diameter does vary with the solar distance. Thus he must specify both a solar distance and a lunar distance at which the apparent diameters of the sun and moon will appear equal. He states a value for the apparent solar diameter when the sun is at apogee (0;31,40°) and a value for the apparent lunar diameter when the moon is at its greatest distance in syzygy (0;30°), without indicating how he arrived at these values. From the apparent solar and lunar diameters and the greatest lunar distance in syzygy, Copernicus computes the lunar distance at which the lunar diameter will appear equal to the diameter of the sun at apogee. This is Copernicus' method in principle. In practice, in his first computation of the solar distance in the manuscript, he neglected to calculate the lunar distance at which the solar and lunar diameters would appear equal and simply used the greatest distance in syzygy.

In the Copernican solar model, the sun actually has a greatest and least distance at apogee. In *De Revolutionibus* iv.19, the chapter on the solar distance, when Copernicus refers to the solar distance at apogee he is speaking of the solar distance in his own time, when the solar eccentricity was near its minimum. In a later chapter, he finds the solar distance in perigee, also for his own time. Reinhold distinguishes between greatest and least solar distances at apogee and perigee and computes all four distances, as well as the length of the shadow axis in all four cases.

The derivation of the solar distance follows the same pattern in all three authors, and I shall here present it in general terms. It assumes an arrangement

[1] Goldstein [1], p. 9.

[2] Rosen [1], p. 162.

of the sun, moon and earth as illustrated in Figure 3.1. The plane of the ecliptic passes through the centers of these three bodies cutting off great circles. ABC is the circle of the sun with center D, NZO is the circle of the moon with center L, and GFE is the circle of the earth with center K. ASC is the intersection of the plane with a cone tangent to the sun and the earth. GES is the portion of this intersection in the shadow cone. QMR is the diameter of the shadow at distance KM, where KM is equal to the lunar distance KL. Since the moon has been placed at the distance where the solar and lunar diameters appear equal, it is possible to form a cone tangent to both the sun and the moon. AKC is the intersection of the plane of the ecliptic with this cone, the apex of which is at the center of the earth. It is assumed that the circles on the spheres of tangency of the cones differ inappreciably from great circles, and thus that the diameters of these circles differ inappreciably from the diameter of the spheres. The four diameters, ADC, NLO, GKE and QMR, are parallel, and NO is produced to P.

The solar distance DK is found as a function of the lunar distance KL and the angular radii of the moon ($\angle LKO$) and the shadow ($\angle MKR$) at this distance, assuming the radius of the earth is one. From the angular radii and the lunar distance the values of LO and MR can be determined. Since LOP and MR are equidistant from and parallel to KE,[3]

$$2KE = 2^{tr} = MR + (LO + OP)$$

and thus OP can be determined. Then in triangles KCE and KCD

$$\frac{1}{OP} = \frac{CK}{CO} = \frac{DK}{DK - KL},$$

and the solar distance is

$$DK = \frac{KL}{1 - OP}.$$

Ptolemy and Copernicus also find the distance of the apex of the shadow cone from the center of the earth, which I refer to as the length of the shadow axis. In triangle KES

$$\frac{1}{MR} = \frac{KS}{KS - KM},$$

and thus

$$KS = \frac{KM}{1 - MR}.$$

1. PTOLEMY'S SOLAR DISTANCE (Table 3.1)

In *Almagest* v.14, Ptolemy found that the angular lunar radius is 0;15,40° when the moon is at greatest distance in syzygy (64;10^{tr}). In triangle KLO, if KO

[3] This is assumed by Ptolemy and Copernicus and proved by Reinhold in his commentary on Ptolemy.

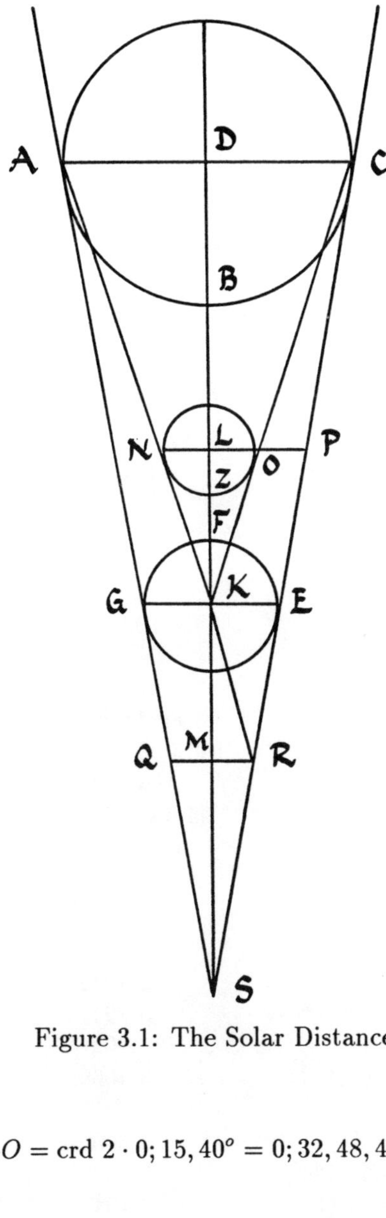

Figure 3.1: The Solar Distance

is 120,

$$LO = \text{crd } 2 \cdot 0;15,40^o = 0;32,48,46,40,$$

and

$$KL = \text{crd } 2 \cdot (90^o - 0;15,40^o) = 119;59,55,28,40.$$

Ptolemy has 0;32,48 for LO, and approximates KL by 120. The true lunar radius is then

$$LO = 0;32,48 \cdot \frac{64;10^{tr}}{120} = 0;17,32,20^{tr}.$$

Ptolemy gives $0;17,33^{tr}$, however. This can be obtained using the more accurate

Table 3.1

PTOLEMY'S SOLAR DISTANCE AND THE LENGTH OF THE SHADOW AXIS

	PTOLEMY	REINHOLD's Recomputation	REINHOLD's values converted to PTOLEMY's norms*
Lunar and solar diameters	0;31,20°	0;31,20°	
Lunar distance at apogee	64;10^{tr}	231 589	64;19,49^{tr}
Shadow-moon radius ratio	13:5	122:47	
∠LKO	0;15,40°	0;15,40°	
∠KOL	89;44,20°	89;44,20°	
if KO	120	100 000 000	[120]
then LO	0;32,48	455 714	[0;32,48,41...]
KL	120	99 998 947	[119;59,55,27...]
if KL	64;10^{tr}	231 589	64;19,49^{tr}
and KE	1^{tr}	3 600	1^{tr}
then LO	0;17,33^{tr}	1 055	0;17,35^{tr}
MR	0;45,38^{tr}	2 738$\frac{1}{2}$	0;45,38$\frac{1}{2}^{tr}$
$LO+MR$	1; 3,11^{tr}	3 794	[1; 3,14^{tr}]
$LP+MR$	2^{tr}	7 200	[2^{tr}]
OP	0;56,49^{tr}	3 406	[0;56,46^{tr}]
if DK	1	3 600	1
then DL	0;56,49	3 406	[0;56,46]
KL	0; 3,11	194	[0; 3,14]

* Numbers in brackets do not appear in Reinhold's text.

Table 3.1 (cont.)

	PTOLEMY	REINHOLD's Recomputation	REINHOLD's values converted to PTOLEMY's norms*
if KL	$64;10^{tr}$	231 589	$64;19,49^{tr}$
and KE	1^{tr}	3 600	1^{tr}
then DK	$1\ 210^{tr}$	4 297 528	$1\ 194^{tr}$
if KE	1^{tr}	3 600	
then MR	$0;45,38^{tr}$	$2\ 738\frac{1}{2}$	
if KS	1		
and $2KS$		7 200	
then MS	0;45,38		
$2MS$		5 477	
KM	0;14,22		
$2KM$		1 723	
if KM	$64;10^{tr}$	231 589	
and KE	1^{tr}	3 600	
then MS	$203;50^{tr}$		
KS	268^{tr}	967 754	269^{tr}

* Numbers in brackets do not appear in Reinhold's text.

value of crd $2 \cdot \angle LKO$ given above:

$$LO = 0;32,48,46,40 \cdot \frac{64;10^{tr}}{120} = 0;17,32,44,57,13,20^{tr}.$$

Multiplying by the ratio of shadow radius to lunar radius, the true shadow radius is

$$MR = \frac{13}{5} \cdot 0;17,33^{tr} = 0;45,38^{tr}.$$[4]

Ptolemy then finds OP. Since LOP, KE and MR are parallel and KL equals KM,

$$LP = LO + OP = 2KE - MR$$

and

$$OP = 2KE - (MR + LO).$$

If KE is 1^{tr}, then

$$OP = 2^{tr} - (0;45,38^{tr} + 0;17,33^{tr})$$

$$= 2^{tr} - 1;3,11^{tr} = 0;56,49^{tr}.$$

In triangles KCE and KCD, with DC, LOP, and KE parallel,

$$\frac{KE}{OP} = \frac{CK}{CO} = \frac{DK}{DL}$$

and

$$DL = DK - KL.$$

If DK is 1, then

$$DL = 1 \cdot \frac{0;56,49^{tr}}{1^{tr}} = 0;56,49,$$

and

$$KL = 1 - 0;56,49 = 0;3,11.$$

Then if KL is $64;10^{tr}$, the solar distance is

$$DK = 64.10^{tr} \cdot \frac{1}{0;3,11} = 1\ 209;25,26^{tr}.$$

Ptolemy gives 1210 earth radii as the solar distance.

In similar triangles KSE and MSR,

$$\frac{KE}{MR} = \frac{KS}{MS}$$

---

[4] Using the true angular radius of the shadow,

$$MR = \frac{\text{crd } 2 \cdot 0;40,40^{o}}{\text{crd } 2 \cdot (90^{o} - 0;40,40^{o})} \cdot 64;10^{tr} = 0;45,33^{tr}.$$

and

$$KM = KS - MS.$$

If KS is 1,

$$MS = 1 \cdot \frac{0;45,38^{tr}}{1^{tr}} = 0;45,38$$

and

$$KM = 1 - 0;45,38 = 0;14,22.$$

Then if KM is 64;10^{tr},

$$MS = 64;10^{tr} \cdot \frac{0;45,38}{0;14,22} = 203;48,53^{tr},$$

for which Ptolemy has 203;50^{tr}. Thus the length of the shadow axis is

$$KS = KM + MS = 64;10^{tr} + 203;50^{tr} = 268^{tr}.$$

It would have been simpler to find KS directly:

$$KS = 64;10^{tr} \cdot \frac{1}{0;14,22} = 267;58,53^{tr}.$$

Ptolemy has 268^{tr} for the distance of the apex of the shadow cone from the center of the earth.

2. REINHOLD'S RECOMPUTATION OF PTOLEMY'S SOLAR DISTANCE (Table 3.1)

Reinhold accepts Ptolemy's values of 0;15,40° for the angular lunar radius and 0;40,40° for the angular shadow radius, but uses his own recomputed value for the greatest lunar distance in syzygy, 64;19,49^{tr}. In order to avoid sexagesimal computations, Reinhold converts 64;19,49^{tr} into 231 589, where 1^{tr} equals 3600. The true lunar radius is

$$LO = KL \tan \angle LKO,$$

and Reinhold finds the equivalent

$$LO = KL \cdot \frac{\sin \angle LKO}{\sin \angle LOK}.$$

Reinhold's eight-place sine values are not correct. He has

$$LO = 231\ 589 \cdot \frac{455\ 714}{99\ 998\ 947},$$

whereas the sine of 0;15,40° should actually be 0.004557233 and the sine of 89;44,20° should be 0.999989616. His values yield 1055 for LO or 0;17,35^{tr}.

Reinhold uses 122:47 for the ratio of shadow to lunar radius, rather than Ptolemy's approximation of 13:5. The true shadow radius is then

$$MR = \frac{122}{47} \cdot 1055 = 2738\ 24/47.$$

Reinhold has $2738\frac{1}{2}$.[5] In subsequent computations he uses both 2739 and $2738\frac{1}{2}$.

In computing OP, Reinhold finds

$$LO + MR = 1055 + 2739 = 3794,$$

and

$$OP = 7200 - 3794 = 3406.$$

In triangles KCE and KCD, with DK equal to 3600,

$$DL = DK \cdot \frac{OP}{KE} = 3600 \cdot \frac{3406}{3600} = 3406,$$

and

$$KL = DK - DL = 3600 - 3406 = 194.$$

Then if KL is 231 589, the solar distance is

$$DK = 231\ 589 \cdot \frac{3600}{194} = 4\ 297\ 528.$$

In terrestrial radii, this is equivalent to 1193.7^{tr} and Reinhold gives 1194^{tr} as the solar distance.

In computing KS, Reinhold does not round MR from $2738\frac{1}{2}$ to 2739, but doubles his values in order to avoid working with the one half. Thus if KS is 3600, $2KS$ is 7200, $2MR$ is 5477, and

$$2MS = 2KS \cdot \frac{2MR}{2KE} = 7200 \cdot \frac{5477}{7200} = 5477.$$

Then

$$2KM = 2KS - 2MS = 7200 - 5477 = 1723.$$

If KM is 231 589, then

$$KS = 231\ 589 \cdot \frac{7200}{1723} = 967\ 754.$$[6]

Converting to terrestrial radii,

$$KS = 268; 49, 14^{tr}.$$

Reinhold gives 269^{tr} for the distance of the apex of the shadow cone from the center of the earth.

5 Use of the angular radius, rather than the ratio, produces a shadow radius

$$MR = 64; 19, 49^{tr} \tan 0; 40, 40^{o} = 0; 45, 40^{tr}.$$

6 Reinhold has first written 967 743 for KS.

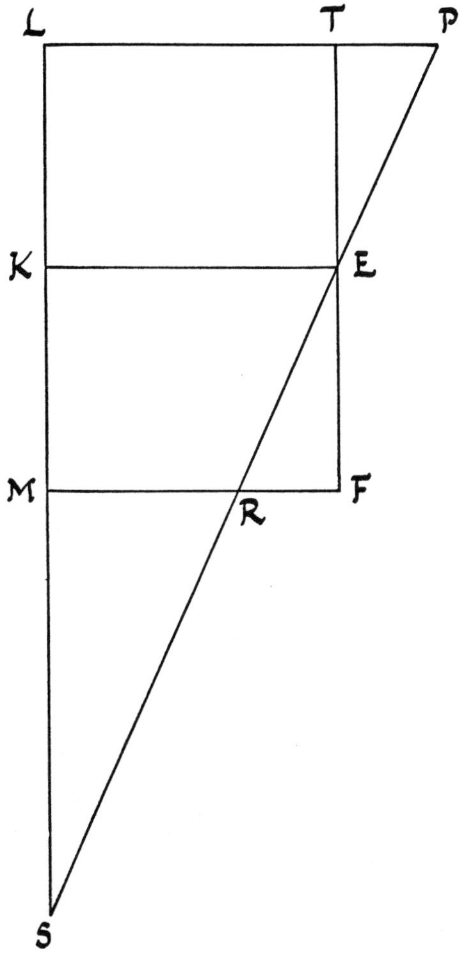

Figure 3.2: Reinhold's Lemma

Following the determination of the two distances, Reinhold states and proves a "lemma." In Figure 3.2, LP, KE and MR are parallel and KL is equal to KM. Reinhold wishes to prove that

$$2KE = MR + LP.$$

TEF is drawn parallel to LKM, meeting MR extended in F, and intersecting SRP in E. Then, in parallelograms $LKET$ and $KMFE$

$$LK = TE$$

and

$$KM = EF. \qquad \text{(Euclid i.34)}$$

Since LK and KM are equal,

$$TE = EF.$$

In the same parallelograms

$$LT = KE = MR \qquad \text{(Euclid i.34)}$$

and thus

$$LT + MF = 2KE.$$

In triangles FER and TEP

$$\angle TEP = \angle FER, \qquad \text{(Euclid i.15)}$$

$$\angle PTE = \angle RFE, \qquad \text{(Euclid i.29)}$$

and

$$TE = EF.$$

Thus the two triangles are congruent and

$$TP = RF.$$

Then

$$MF = MR + RF = MR + TP$$

and

$$LT = LP - TP.$$

Adding the two equations

$$MF + LT = 2KE = MR + LP. \qquad \text{Q.E.O.}$$

Following the lemma is a rather lengthy discussion of the shape of the shadow cast by the earth, in which Reinhold considers why the shadow is conical and under what conditions the shadow would assume a different form. This section of the manuscript is unusual in that Reinhold's commentary is normally limited to the contents of the *Almagest*, and for this reason I translate it in full.[7]

---

[7] References to the lettering in Reinhold's diagram have been changed throughout to agree with Figure 3.1. The diagrams in the *Almagest* and in Reinhold's commentary on Ptolemy are similarly lettered (in Greek), as are the diagrams in *De Revolutionibus* and Reinhold's commentary on Copernicus. It is the latter which has been reproduced in Figure 3.1 and which is used here for both Ptolemy and Copernicus in order to facilitate comparison and avoid confusion. Some of the numbering of the paragraphs has been changed in the translation, for the sake of clarity, but Reinhold's numbering and lettering (transcribed from Greek to Roman) are preserved in the transcription.

Cur Umbra Terrae Sit *Konoeides*

Quoniam in precedenti demonstrationi ostensum est quod qualium semidiameter terrae est 3600″ particularum scilicet MN, talium semidiameter PR umbrae $2738\frac{1}{2}''$, manifestum est umbram terrae nec esse *kylindroeide* nec *kalathoeide*, sed *konoeide*, hoc est, paulatim descrescere atque attenuari donec in mucronem desinat, quemadmodum et finis totius demonstrationis patefecit. Est autem haec proportio semidiametri terrae et umbrae necessario coniuncta velut comes, primum distantiae maximae lunae a terris quam diximus esse 64 1/3 semidiametrorum terrae fere, deinde proportioni semidiametri lunae et umbrae et caeteram.

Caeterum quae de proportionibus trium corporum solis, lunae, et terrae, aut umbrae forma, in genere dici possunt hoc ordine nobis patent.

1. Patet ex praecedenti demonstratione semidiametrum umbrae esse fere dodrantem semidiametri terrae.

2. Lunae autem semidiametrum paulo plus quadrante.

3. Igitur necessario umbra terrae *konoeides* existit quippe quae subinde fit tenuior.

4. Sol igitur maior terra quia opacum corpus non proiicit umbram *konoeide* nisi luminoso corpore maiori existente.

5. Et luna minor terra quoniam ipsius diameter paulo maior est quam quarta pars diametri terrae.

6. Ideoque eadem luna multo minor sole existit.

Quod si cui libeat cogitare quomodo umbra terrae existere potuerit vel *kylindroeides* vel *kalathoeides*, is experietur distantiae lunari et proportioni semidiametrorum umbrae et lunae, necessario obnoxiam esse formam umbrae ita ut si alterutra mutetur necessario et species umbrae mutetur.

I. Si mutetur distantia lunaris.
Nam si aliam lunae distantiam a terris accipias, non mutata interim proportione semidiametrorum lunae atque umbrae, sequetur omnino umbrae variatio aliaque trium corporum proportio.

1. Ut si sumas maximam lunae a terris distantiam 84 semidiametrorum terrae, erit semidiameter quidem lunae paulo minor quam duae quintae unius semidiametri terrae, semidiameter autem umbrae omnino par semidiametro terrae. Proinde et umbra terrae existeret *kylindroeides*, et luminosum corpus solis non maius terra iuxta regulas notissimas opticorum.

II. Sed si maximam illam lunae remotionem a terris maiorem sumas quam 84 semidiametrorum terrae, necessario umbra terrae evadet *kalathoeides*, ac solis corpus magnitudine superabitur a terra.

Idque varietate infinita potuisset accidere quod paucis exemplis seu casibus aestimare licet.

1. Ut si ponatur lunae distantia 170 semidiametrorum terrae fere, erit semidiameter quidem lunae paulo maior quam dodrans semidiametri umbrae terrae. At semidiameter umbrae dupla semidiametri terrae.

2. Quod si eadem distantia intelligatur 220 semidiametrorum terrae, lunaris quidem diameter aequabitur diametro terrae, at umbrae diameter maior erit quam duplo diametro terrae tribus scilicet quintis unius diametri fere.

3. Sin autem statuas maximam lunae a terris distantiam 254 semidiametrorum terrae, erit semidiameter umbrae tripla semidiametro terrae, semidiameter autem lunae maior semidiametro terrae sexta fere parte unius.

II. Similis umbrae varietas comitabitur mutationem proportionis semidiametrorum lunae et umbrae etiamsi eadem lunae distantia retineatur.

1. Nam si proportio semidiametri lunae ad semidiametrum umbrae fuerit sicut 1055 ad 3600, hoc est, sicut semidiameter lunae ad semidiametrum terrae ut constat ex praecedentibus, erunt aequales semidiametri terrae atque umbrae per 9 quinti Euclidis, et umbra terrae *kylindroeides*.

II. Quod si semidiameter lunae ad semidiametrum umbrae minorem rationem habeat quam 1055 ad 3600, erit, per 10 quinti, semidiameter umbrae maior semidiametro terrae, ac propterea umbra ipsa specie existet *kalathoeidei*.

1. Ut si proportio semidiametri lunae ad semidiametrum umbrae fuerit sicut 1055 ad 5400, erit umbrae semidiameter sesquialtera semidiametro terrae.

2. Si fuerit sicut 1055 ad 7200, erit umbrae semidiameter dupla semidiametro terrae.

3. Si fuerit sicut 1055 ad 10800, erit tripla proportio.

4. Sed si fuerit sicut 1055 ad 14400, erit quadrupla proportio.

Et sic deinceps multiplicabitur umbra in infinitum, quanto minor fuerit proportio semidiametri lunae ad semidiametrum umbrae. (Ms.R.181v-183r)

Why the Shadow of the Earth is Conical

Since in the preceding demonstration it was shown that when the radius of the earth, namely KE, is 3600, the radius of the shadow, MR, is $2738\frac{1}{2}$, it is obvious that the shadow of the earth is neither cylindrical nor an inverted cone [*kalathoeide* = basket-shaped] but conical, i.e., it gradually decreases and thins until it terminates in a sharp point, as the end of the whole demonstration disclosed. However, this relation of the radii of the earth and shadow is necessarily joined as a companion, first, to the greatest distance of the moon from the earth, which we said was about 64 1/3 earth radii, second, to the ratio of the radii of the moon and shadow, et cetera.

On the other hand, those things concerning the relation of the three bodies of the sun, moon and earth, or the shape of the shadow, can be said to be accessible to us in this order.

1. It is obvious from the previous demonstration that the radius of the shadow is approximately three quarters the radius of the earth.

2. On the other hand, the radius of the moon is a little more than a quarter.

3. Therefore the shadow of the earth necessarily appears conical, since immediately afterwards it is smaller.

4. Therefore the sun is greater than the earth, because an opaque body cannot project a conical shadow unless the illuminating body is larger.

5. And the moon is smaller than the earth since its diameter is a little greater than a quarter of the diameter of the earth.

6. Thus the moon is much smaller than the sun.

But if one would like to know how the shadow of the earth could be either cylindrical or an inverted cone, this can be learned by varying the distance to the moon and the ratio of the radii of the shadow and moon, for if either of these is changed, the shape of the shadow is changed.

I. If the lunar distance is changed.

For if a different distance of the moon from the earth is accepted, meanwhile not changing the ratio of the radii of the moon and shadow, it follows that the relation of the shadow to the other three bodies will be altogether different.

1. If the greatest distance of the moon from the earth is assumed to be 84 earth radii, the radius of the moon will be a little less than two fifths of one earth radius, and the radius of the shadow will be equal to the radius of the earth. Thus the shadow of the earth would appear cylindrical, and the illuminating body of the sun could not be greater than the earth according to the well known laws of optics.

2. But if the greatest distance of the moon from the earth is assumed to be greater than 84 earth radii, the shadow of the earth necessarily becomes an inverted cone, and the body of the sun is surpassed in size by the earth. Thus an unlimited variety of possibilities can occur, and one can get some idea of this from a few examples.

i. If the lunar distance is placed at approximately 170 earth radii, the lunar radius will be a little greater than three quarters of the radius of the earth's shadow. But the radius of the shadow will be twice the radius of the earth.

ii. If the same distance is understood to be 220 earth radii, the diameter of the moon will be equal to the diameter of the earth, but the diameter'of the shadow will be greater than twice the diameter of the earth, namely, about two and three fifths of the diameter.

iii. But if the greatest distance of the moon from the earth is 254 earth radii, the radius of the shadow will be three times the radius of the earth, and the radius of the moon will be greater than the radius of the earth by about a sixth.

II. Similarly the shadow will be altered by changing the ratio of the lunar and shadow radii while leaving the lunar distance fixed.

1. For if the ratio of the lunar to shadow radii were 1055 to 3600, i.e., the same as the radius of the moon to the radius of the earth as established in the preceding, the radius of the earth would be equal to that of the shadow, by Euclid v.9, and the shadow of the earth would be cylindrical.

2. But if the radius of the moon to the radius of the shadow had a ratio of less than 1055 to 3600, then, by Euclid v.10, the radius of the shadow will be greater than the radius of the earth, and the shadow itself will have the shape of an inverted cone.

i. If the ratio of the radius of the moon to the radius of the shadow is as 1055 to 5400, then the radius of the shadow will be one and one half times the radius of the earth.

ii. If it is as 1055 to 7200, the radius of the shadow will be twice the radius of the earth.

iii. If it is as 1055 to 10800, the ratio will be three.

iv. But if it is as 1055 to 14400, the ratio will be four.

And in this way as the shadow is increased in regular order to infinity, by so much less will the ratio of the radius of the moon be to the radius of the shadow.

The solar distance is extremely sensitive to small changes in the lunar distance and the moon-shadow ratio. Swerdlow[8] has determined that if the ratio is fixed at 5:13, the solar distance becomes infinite when the lunar distance is $60;57^{tr}$, and the solar and lunar distances become equal when the lunar distance is $127;54^{tr}$. If Ptolemy, like Hipparchus, had assumed that the solar and lunar diameters appeared equal when the moon was at its mean distance in syzygy (59^{tr} for Ptolemy), he would not have been able to obtain the solar distance using Hipparchus' geometric method.

In the passage quoted above, I believe Reinhold is not interested in the changes in the solar distance produced by variations in the parameters. He is

[8] Swerdlow [1], p. 65.

solely concerned with the shape of the shadow. The moon shadow ratio has been defined as the ratio of the apparent diameters of the moon and the shadow at that distance where the lunar and solar diameters appear equal. Thus the ratio varies as a function of the solar and lunar distances. The definition assumes, however, that the shape of the shadow is a cone and that the moon and sun have fixed true diameters. When Reinhold changes the shape of the shadow while holding the moon-shadow ratio constant, the ratio is no longer simply a function of the solar and lunar distances.

In the first case which Reinhold describes, where the lunar distance varies while the moon-shadow ratio is fixed, Reinhold uses Ptolemy's approximate value of 5:13 for the ratio, though he had used 122:47 in his redetermination of the Ptolemaic solar distance. The first example determines the limiting condition which produces a cylindrical shape of the shadow. This will occur when the shadow radius is 1^{tr}, in which case the lunar radius is $5/13^{tr}$.[9] Reinhold is still assuming that the apparent lunar radius is 0;15,40°. The lunar distance under these conditions is then

$$KL = \frac{5^{tr}}{13} \cot 0;15,40^{o} = 84.4^{tr}.$$

Of course the solar radius itself must be 1^{tr} in order for the shadow to be cylindrical. If the apparent solar diameter is still 0;15,40°, then the solar distance would be reduced to

$$KD = 1^{tr} \cot 0;15,40^{o} = 220^{tr}$$

if the shadow were cylindrical.

If the lunar distance is greater than 84^{tr}, the shape of the shadow will be an inverted cone, and the solar radius will be less than the terrestrial radius. Reinhold gives three examples. In the first, the shadow radius is 2^{tr}. Thus

$$R_m = \frac{5}{13} \cdot 2^{tr} = \frac{10^{tr}}{13} = \frac{40^{tr}}{52},$$

which Reinhold says is a little greater than 3/4 or 39/52. The lunar distance is then

$$KL = \frac{10}{13} \cdot \cot 0;15,40^{o} = 168.79^{tr}.$$

Reinhold says this is approximately 170^{tr}.

If the sun were a point source of light, then a shadow radius of 2^{tr} would imply that the solar and lunar distances were equal. Since the sun has an actual diameter, a shadow radius of 2^{tr} implies that the sun must be closer to the earth than the moon. This applies equally to the last two examples. In the first, the shadow radius is $13/5^{tr}$ or $2\ 3/5^{tr}$. The lunar radius is thus

$$R_m = \frac{5}{13} \cdot \frac{13^{tr}}{5} = 1^{tr},$$

[9] $5/13^{tr}$ is, as Reinhold says, a little less than $2/5^{tr}$ since 5/13 is equal to 50/130, whereas 2/5 is equal to 52/130.

and the lunar distance is

$$KL = \cot 0;15,40^{o} = 219.43^{tr},$$

which Reinhold says is approximately 220^{tr}. In the last example the shadow radius is 3^{tr}. Thus the lunar radius is

$$R_m = \frac{5}{13} \cdot 3^{tr} = \frac{15^{tr}}{13}.$$

Reinhold says this is a little greater than $7/6^{tr}$, but actually it is less, since 15/13 is equivalent to 90/78, and 7/6 is equivalent to 91/78. The lunar distance is then

$$KL = \frac{15}{13} \cdot \cot 0;15,40^{o} = 253.19^{tr},$$

which Reinhold approximates as 254^{tr}.

In the second set of examples, the lunar distance is fixed while the ratio is allowed to vary. This case is of particular interest because in his commentary on *De Revolutionibus* Reinhold uses this technique to arrive at a solar distance which he considers suitable. There, of course, the assumption is that the shadow is conical. Here, Reinhold is interested in other shapes of the shadow. In his recomputation of Ptolemy's solar distance, he derived, from a lunar distance of $64;19,49^{tr}$ and an angular lunar radius of $0;15,40^{o}$, a true lunar radius of $0;17,35^{tr}$ or 1055 (where 1^{tr} equals 3600). The shape of the shadow will be a cylinder only if the terrestrial and shadow radii are equal. This will be the case if the ratio is 1055:3600, for then the shadow radius is 1^{tr}. If the ratio is less than 1055:3600, the radius of the shadow will be greater than that of the earth. Reinhold gives four examples. In the first, the shadow radius is $1\frac{1}{2}^{tr}$ or 5400, and thus the ratio of lunar to shadow radius is 1055:5400. In the second, the shadow radius is 2^{tr} or 7200 and the ratio is 1055:7200. In the third, the shadow radius is 3^{tr} or 10800 and the ratio is 1055:10800. Finally, in the fourth, the shadow radius is 4^{tr} or 14400 and the ratio is 1055:14400. These are more than enough examples to illustrate Reinhold's point.

Assuming the apparent solar radius is still $0;15,40^{o}$, the solar distance in these examples varies as it did in the first set, namely, if the shadow radius is 1^{tr}, the solar distance is 220^{tr}, and if it is greater than or equal to 2^{tr}, then the solar distance is less than the lunar distance.

3. COPERNICUS' SOLAR DISTANCE

Copernicus begins his chapter on the solar distance with a summary of Ptolemy's method and results. This summary follows Ptolemy exactly and the numerical results which Copernicus states do not differ from those of Ptolemy. In his description of the Ptolemaic derivation Copernicus specifies that the earth is at its greatest distance from the sun. Thus Copernicus assumes that

Ptolemy's solar distance of 1210^{tr} represents the greatest, not the mean, solar distance.[10]

The discussion of Ptolemy is followed by a summary of al-Battānī's results for the solar distance. Al-Battānī accepted Ptolemy's values of $64;10^{tr}$ for the greatest distance of the new and full moon, 0;31,20° for the angular diameter of the sun at apogee, and 13:5 for the ratio of the diameter of the shadow to the diameter of the moon at the distance $64;10^{tr}$. As Copernicus states, however, al-Battānī assumed that the diameter of the moon at this distance was 0;29,30°. Thus the diameter of the shadow is 1;16,45°.[11] Al-Battānī's results are 1146^{tr} for the distance of the sun at apogee, and 254^{tr} for the axis of the shadow.[12] Copernicus' remark on al-Battānī's parameters and results is that "they cannot be put together by any computation" ("tamen nulla ratione possunt coniungi"). This is quite true.[13] There are many difficulties in al-Battānī's determination of the solar distance. While ostensibly using Ptolemy's geometric method, he actually determines the solar distance simply by multiplying his own lunar distance by the ratio of Ptolemy's solar to lunar distance.

It is of interest to note a similarity in the methods of Copernicus and al-Battānī. Both astronomers find the diameter of the sun at greatest solar distance and subsequently compute the lunar distance at which the diameter of the moon will appear equal to this minimum solar diameter. Thus Copernicus and al-Battānī allow for the possibility of annular solar eclipses. Like Ptolemy, they assume that the ratio of the lunar and shadow diameters applies to both the angular radius and its tangent. They also assume that the ratio remains valid when the lunar distance is changed, namely from greatest distance in syzygy to that distance at which solar and lunar diameters appear equal.

The last section of *De Revolutionibus* iv.19 contains Copernicus' own determination of the solar distance. It is full of difficulties and resists a complete and consistent explanation. The source of the parameters used by Copernicus

---

[10] There are very few errors, or changes, in this section of Copernicus' manuscript. The value 0;56,49 for OP occurs three times. The second time it appears, the 6 has inexplicably been changed to an 8. Copernicus first writes for the length of the axis of the shadow (KS) Ptolemy's value for MS, 203;50.

[11] Accurately

$$D_{sh} = \frac{13}{5} \cdot 0;29,30^o = 1;16,42^o.$$

Copernicus states 1;16,45°. Al-Battānī actually gives

$$R_{sh} = 0;38,20^o,$$

so that

$$D_{sh} = 1;16,40^o,$$

which in *Epitome* v.21 is rounded to 1;17°.

[12] Al-Battānī actually has 254 $2/3^{tr}$, but Copernicus and Regiomontanus give 254^{tr}.

[13] Cf. Nallino [1], v.1, p. 61. There is a detailed analysis of al-Battānī's determination of the lunar and shadow radii, mainly by Schiaparelli, on pp. 230-237 of this volume. Al-Battāni ¯'s determination of the solar distance is discussed in detail in Swerdlow [3].

is unclear. He begins the computation with one set of parameters, some of which are still in use in later chapters, but he subsequently returned to this chapter and changed his values, which forced him to alter his results. Some of the inconsistencies can only be explained as computational errors.

Copernicus' values for the angular diameters of the sun and moon, and for the ratio of the lunar and shadow radii differ from Ptolemy's. His angular diameter of the sun in apogee is 0;31,40°. The corresponding quantity for al-Battānī was 0;31,20°, which was also Ptolemy's value, although for Ptolemy this in fact applied to the mean distance of the sun. Copernicus remarks that the angular diameter of the sun should be somewhat larger now than it was before the time of Ptolemy. According to Copernican solar theory, the eccentricity of the model changes slowly with time. The eccentricity was at a maximum 64 years before Christ, and in Copernicus' time is near the minimum (*De Revolutionibus* iii.20 and 21). The distance of the sun at apogee is thus less in Copernicus' time than it was in Ptolemy's time and, in making the solar diameter somewhat larger, Copernicus is proceeding according to his theory.[14] The angular diameter of the moon at its greatest distance in syzygy is 0;30° for Copernicus, whereas Ptolemy believed the value at this distance was the same as the solar diameter, 0;31,20°, and al-Battānī found 0;29,30°.[15]

Before finding the solar distance DK and length of the axis of the shadow KS, Copernicus gives his results for the quantities LO, MR and OP. There are two sets of values for these quantities (see Table 3.2). The first set is computed from a lunar distance of 65;30^{tr} and a ratio of shadow radius to lunar radius of 79:30. This lunar distance is Copernicus' value for the greatest distance of the moon in syzygy. The ratio 79:30 follows from Ptolemy's value for the angular diameter of the shadow, 1;21,20°, adjusted by the ratio of Ptolemy's

---

[14] Copernicus' value of 0;31,40° is not unreasonable. If the solar diameter was 0;31,20° when the sun was at its maximum distance in apogee, 10 417, and in Copernicus' time the solar distance is 10 323, then

$$\frac{10\,417}{10\,323} = \frac{\sin \angle AKC/2}{\sin 0;31,20^o/2},$$

and the solar diameter is approximately 0;31,34°. A direct comparison yields

$$\angle AKC = \frac{10\,417}{10\,323} \cdot 0;31,20^o = 0;31,37^o.$$

Ptolemy's value for the lunar and solar diameters, however, was determined from Babylonian eclipse observations made some six to seven centuries before Christ.

[15] Al-Battānī's value is for a lunar distance of 64;10^{tr}, and for Copernicus the lunar distance is 65;30^{tr}. If

$$\frac{64;10}{65;30} = \frac{\angle NKO/2}{0;29,30^o/2},$$

the lunar diameter is 0;30,7°.

and Copernicus' lunar distances, i.e.,

$$\frac{64;10}{65;30} \cdot 1;21,20^o = 1;19,3,24 = 1;19^o,$$

so that

$$\frac{1;19^o}{0;30^o} = \frac{79}{30}.$$

Copernicus says that the ratio in his time should be a little larger than the previous value, i.e., Ptolemy's value of 13:5.

In his first calculation of the solar distance, i.e., in the original manuscript values, it appears that Copernicus was not really thinking about what he was doing. Since the angular solar diameter is 0;31,40° and the angular lunar diameter is 0;30°, the two are not equal when the moon is at its greatest distance in syzygy. Copernicus ignores this and computes with a lunar distance of 65;30^{tr}. His true radius of the moon, LO, should then have been

$$LO = 65;30^{tr} \sin 0;15^o$$

or perhaps

$$LO = 65;30^{tr} \sin 0;15,50^o.$$[16]

Copernicus finds 0;18,11^{tr} for LO. This corresponds to no reasonable combination of the parameters.

Copernicus next computes MR, the true shadow radius, which should be

$$MR = \frac{79}{30} \cdot 0;18,11 = 0;47,53^{tr}.$$

Copernicus has 0;45,10^{tr}. Now

$$0;45,10^{tr} = \frac{79}{30} \cdot 0;17,9^{tr},$$

and the value 0;17,9 for LO does occur in the manuscript.[17] Finally, Copernicus' value of OP is correctly computed:

$$\begin{aligned} LOP &= 2KE - MR \\ &= 2^{tr} - 0;45,10^{tr} = 1;14,50^{tr}, \end{aligned}$$

and

$$\begin{aligned} OP &= LOP - LO \\ &= 1;14,50^{tr} - 0;18,11^{tr} = 0;56,39^{tr}. \end{aligned}$$

[16] Technically one should use the tangent but the difference for such small angles is inappreciable.

[17] Judging from the way the corrections are made in the manuscript, it is possible that 0;18,11 was corrected to 0;17,9, which was then used to compute MR. However, the value 0;18,11 for LO is used subsequently to compute OP. Additional confusion arises from Copernicus' value for LO computed with a different lunar distance and moon-shadow radius ratio, namely 0;17,8.

Table 3.2

COPERNICUS' SOLAR DISTANCE AND THE LENGTH OF THE SHADOW AXIS

	COPERNICUS First version in manuscript	COPERNICUS Changes in Manuscript	REINHOLD's Recomputation
∠LKO			0;15°
∠KOL			89;45°
if KO			100 000 000
then KL			99 999 048
LO			436 321
if KL			64;57^{tr}
and KE			1^{tr}
then LO			0;17^{tr}
Solar diameter at apogee	0;31,40°		0;31,20°
Lunar diameter at apogee	0;30°		0;30°
Lunar distance for equal radii	65;30^{tr}	62^{tr}	62;10^{tr}
Shadow-moon radius ratio	79:30	403:150	403:150
if KO	1^{tr}		0;60^{tr}
then LO	0;18,11^{tr}	0;17,8^{tr}	0;17^{tr}
MR	0;45,10^{tr}	0;46, 1^{tr}	0;45,52^{tr}
LP			1;14, 8^{tr}
OP	0;56,39^{tr}	0;56,51^{tr}	0;57, 8^{tr}
if DK			60
then KL			0; 2,52
if KL			62;10^{tr}
then DK	1579^{tr}	1179^{tr}	1301^{tr}
$KE-MR$			0;14, 8^{tr}
MR			0;45,52^{tr}
if KS			60
then KM			0;14, 8
if KM			62;10^{tr}
then KS	265^{tr}		264^{tr}

The second set of values found in the manuscript agrees with those which appear in the Nuremberg edition. Sometime after completing this chapter on the solar distance,[18] it appears Copernicus realized he had not used the lunar distance at which the solar and lunar diameters appear equal. Like al-Battānī, he now finds the lunar distance at which the apparent lunar diameter is equal to the apparent solar diameter in apogee, in this case 0;31,40°. Al-Battānī's method was unnecessarily indirect. The distance can be found quite simply by taking

$$65;30^{tr} \cdot \frac{\sin 0;15^o}{\sin 0;15,50^o} = 62;0,30^{tr}.$$

Copernicus says that the sun is not covered by the moon unless the lunar distance is less than 62^{tr}. This is the value he uses in subsequent computations.

Copernicus also changes the ratio of the radii of the moon and shadow, from 79:30 to 403:150. Neugebauer[19] has suggested that this new ratio is an adaptation of Ptolemy's value of 13:5. For Ptolemy, solar and lunar diameters appear equal at a lunar distance of $64;10^{tr}$, whereas for Copernicus this is true at 62^{tr}. Now

$$\frac{64;10}{62} = \frac{3850}{3720} \cong \frac{31}{30},$$

and

$$\frac{13}{5} \cdot \frac{31}{30} = \frac{403}{150}.$$

This modification of the Ptolemaic ratio has the advantage of producing a solar distance on the same order of magnitude as Ptolemy's.

An angular lunar diameter of 0;30° and a ratio of 403:150 imply

$$MR = \frac{403}{150} \cdot 0;30^o = 0;80,36^o.$$

This is Copernicus' value for the angular diameter of the shadow. His value for LO is $0;17,9^{tr}$, which is subsequently changed in the manuscript to $0;17,8^{tr}$. This is derived from

$$LO = 62^{tr} \sin 0;15,50^o.^{20}$$

[18] As late as chapter 23, Copernicus is still using the original manuscript values found in chapter 19.

[19] Neugebauer [2], p. 101.

[20] Accurately computed

$$\sin 0;15,50^o = 0;0,16,34,49,52,...,$$

producing

$$LO = 62^{tr} \cdot 0;0,16,35 = 0;17,8.$$

But note that

$$65;30^{tr} \sin 0;15^o = 65;30^{tr} \cdot 0.00436 = 0;17,8,5...^{tr}.$$

However, if Copernicus was using his table of chords,

The apparent shadow radius, MR, should be

$$MR = 0;17,8^{tr} \cdot \frac{403}{150} = 0;46,1,53^{tr}.$$

Copernicus has 0;46,1^{tr}. His value of OP follows correctly from MR and LO:

$$\begin{aligned} LOP &= 2KE - MR \\ &= 2^{tr} - 0;46,1^{tr} = 1;13,59^{tr}, \end{aligned}$$

and

$$\begin{aligned} OP &= LOP - LO \\ &= 1;13,59^{tr} - 0;17,8^{tr} = 0;56,51^{tr}. \end{aligned}$$

Copernicus' value for the maximum solar distance is 1179^{tr}.[21] This does not follow exactly from either his first or second set of values. In the first case

$$DK = \frac{KL}{1 - OP} = \frac{65;30}{1 - 0;56,39} = 1173;8,3^{tr},$$

and in the second case

$$DK = \frac{62}{1 - 0;56,51} = 1180;57,8^{tr}.$$

If, however, one rounds Copernicus' *initial* value of OP to 0;56,40, then

$$DK = \frac{65;30}{1 - 0;56,40} = 1179;0^{tr}$$

exactly. If this is indeed how Copernicus obtained his value, it would imply that he did not bother to recompute the solar distance after changing his parameters.

Copernicus' value for the length of the axis of the shadow tends to confirm this hypothesis. With the first set of parameters

$$KS = \frac{KL}{1 - MR} = \frac{65;30}{1 - 0;45,10} = 264;56^{tr},$$

$$\begin{aligned} \frac{1}{2}\text{crd } 2 \cdot 0;15,50^o &= 0.00460\ \ 2/3 \\ &= 0.00461 = 0;0,16,35,45,36, \end{aligned}$$

and

$$LO = 62^{tr} \cdot 0;0,16,36 = 0;17,9.$$

[21] Copernicus has first written MDLXXIX, i.e., 1579, but this is totally inexplicable.

but with the second

$$KS = \frac{62}{1 - 0;46,1} = 266;1,51^{tr}.$$

Copernicus' value is 265^{tr}, thus following from the old parameters. It is not at all unusual for Copernicus to make corrections in various steps of his work while leaving the final result untouched and incorrect.[22] Presumably he feels in each case that the final result would not be much altered by carrying through the corrections. His solar distance is not unreasonable, given that he expected a result similar to Ptolemy's. As the culmination of five chapters of theory and computation, however, it is a disappointing conclusion.

4. REINHOLD'S RECOMPUTATION OF THE SOLAR DISTANCE

Reinhold's commentary on *De Revolutionibus* iv.19 is quite unusual. It appears he is satisfied neither with Copernicus' assumptions nor with his own recomputed results. Like his discussion of the shape of the shadow, this is a section of his commentary which shows originality, and I shall describe it in some detail. In summary, the contents of Reinhold's commentary on chapter 19 are as follows:

1. A determination of the lunar distance at which the solar and lunar diameters appear equal
2. A determination of the solar distance under the assumption of a shadow-moon ratio of 13:5 (resulting in $DK = 3108\ 1/3^{tr}$)
3. A determination of the solar distance under the assumption of a shadow-moon ratio of 403:150 (resulting in $DK = 1301^{tr}$)
4. A deleted version of the same determination with a new value for the ratio, presumably Reinhold's own, 19:7 (resulting in $DK = 1187^{tr}$)
5. An analysis which derives the ratio assuming a solar distance of 1200^{tr} (resulting in a ratio of 122:45)
6. A synthesis based on the preceding, where the ratio 122:45 yields a solar distance of 1208^{tr}
7. A determination of the solar distance for maximum and minimum eccentricity of the sun in apogee and perigee
8. A determination of the angular radius of the sun and the lunar distance at which the solar and lunar diameters appear equal for minimum eccentricity of the sun in apogee
9. The length of the axis of the shadow for minimum and maximum eccentricity of the sun in apogee and perigee
10. A verification of the validity of the ratio 122:45 for minimum eccentricity of the sun in apogee.

[22] See, for example, his computation of the first zenith distance observation, and also the computation of the true lunar distance for the second observation.

Reinhold accepts Copernicus' value for the angular diameter of the full moon in apogee, 0;3°, but for a distance of 64;57^{tr}, not 65;30^{tr}. Nowhere does he mention Copernicus' value of 0;31,40° for the angular diameter of the sun in apogee. He consistently restricts himself to Ptolemy's value of 0;31,20° in computing the solar distance. Following Copernican theory, Reinhold believed that the solar eccentricity was at a maximum near the time of Hipparchus and Ptolemy, and that in his own time it was near the minimum. Thus Ptolemy's angular diameter of 0;31,20° applies to maximum eccentricity in apogee, and Copernicus' 0;31,40° to minimum eccentricity. This is essentially the situation Reinhold tries to duplicate. He has a choice between 1) assuming 0;31,20° for maximum eccentricity in apogee and then finding the solar diameter for minimum eccentricity based on this assumption, or 2) assuming 0;31,40 for minimum eccentricity in apogee and then finding the solar diameter for maximum eccentricity. He chooses the former alternative and finds 0;30,36° for the angular diameter of the sun at minimum eccentricity. Reinhold's motivation is clear although he is not explicit, possibly in order to avoid contradicting Copernicus.

The lunar distance at which the solar and lunar diameters appear equal

Reinhold found a greatest lunar distance in syzygy of 64;57^{tr} in his commentary on *De Revolutionibus* iv.17. Assuming an angular lunar diameter of 0;30°, the true lunar radius is

$$LO = 64;57^{tr} \tan 0;15^{o}.$$

Reinhold finds[23]

$$\begin{aligned} LO &= 64;57^{tr} \cdot \frac{\sin 0;15^{o}}{\sin(90^{o} - 0;15^{o})} \\ &= 64;57^{tr} \cdot \frac{436321}{99\,999\,048} = 0;17^{tr}. \end{aligned}$$

The lunar distance at which the solar and lunar diameters will both appear to be 0;31,20° is

$$KL = 0;17^{tr} \cot 0;15,40^{o}.$$

Reinhold finds[24]

$$\begin{aligned} KL &= 0;17^{tr} \cdot \frac{\sin(90^{o} - 0;15,40^{o})}{\sin 0;15,40^{o}} \\ &= \frac{99\,998\,961}{455\,713} = 62;10^{tr}. \end{aligned}$$

These two values, 0;17^{tr} for the true lunar radius and 62;10^{tr} for the lunar distance at which solar and lunar diameters appear equal, are used in all four of

23 The sine of (90° - 0;15°) is correct, but the sine of 0;15° is 4 363 309.

24 Again the sine of (90° - 0;15,40°) is correct but the sine of 0;15,40° is 455 723.

Reinhold's solar distance determinations. Before beginning the computation, he states explicitly that these values are for Ptolemy's time, although he incorrectly gives the apparent solar diameter as 0;30,20°. The numerical values which Reinhol obtains in his four computations are listed in Table 3.3.

The solar distance assuming a shadow-moon ratio of 13:5

Reinhold first computes the solar distance with Ptolemy's value of the ratio. The true shadow radius is

$$MR = 0;17^{tr} \cdot \frac{13}{5} = 0;44,12^{tr}.$$

He correctly computes the value of OP:

$$\begin{aligned} LOP &= 2KE - MR \\ &= 2^{tr} - 0;44,12^{tr} = 1;15,48^{tr}, \end{aligned}$$

and

$$\begin{aligned} OP &= LOP - LO \\ &= 1;15,48^{tr} - 0;17^{tr} = 0;58,48^{tr}. \end{aligned}$$

The solar distance is then[25]

$$DK = \frac{62;10^{tr}}{1^{tr} - 0;58,48^{tr}} = 3108;20^{tr}.$$

This, Reinhold says, is too great a distance, and Copernicus was therefore correct in assuming a different value for the ratio.

The solar distance assuming a shadow-moon ratio of 403:150

Reinhold then begins a new section in the manuscript, entitled *Allos* (in another way). Assuming Copernicus' value of the ratio,[26] the true shadow radius should be

$$MR = 0;17^{tr} \cdot \frac{403}{150} = 0;45,4024^{tr}.$$

Reinhold's value, however, is 0;45,52^{tr}. This, most likely, is a computational error. The rest of the calculation is correct:

$$LOP = 2^{tr} - 0;45,52^{tr} = 1;14,8^{tr},$$

$$OP = 1;14,8^{tr} - 0;17^{tr} = 0;57,8^{tr},$$

[25] It appears that Reinhold has written 2/3, rather than 1/3, for the fractional part.

[26] Reinhold, of course, did not know that Copernicus originally assumed 79:30 for the ratio.

Table 3.3

REINHOLD'S DERIVATION OF THE SOLAR DISTANCE

Solar diameter at apogee	$0;31,20^{o}$
Lunar diameter at apogee	$0;30^{o}$
Lunar distance for equal radii	$62;10^{tr}$
True lunar radius	$0;17^{tr}$

		1	2	3	4 Analysis	5 Synthesis
Shadow–moon radius ratio		13:5	403:150	19:7	122:45	122:45
if	*KE*	$0;60^{tr}$	$0;60^{tr}$	$0;60^{tr}$	$0;60^{tr}$	$0;60^{tr}$
and	*LO*	$0;17^{tr}$	$0;17^{tr}$	$0;17^{tr}$	$0;17^{tr}$	$0;17^{tr}$
then	*MR*	$0;44,12^{tr}$	$0;45,52^{tr}$	$0;46\ 1/7^{tr}$	$0;46,\ 5,40^{tr}$	$0;46,\ 5,20^{tr}$
	LP	$1;15,48^{tr}$	$1;14,\ 8^{tr}$	$1;13\ 6/7^{tr}$	$1;13,54,20^{tr}$	$1;13,54,40^{tr}$
	OP	$0;58,48^{tr}$	$0;57,\ 8^{tr}$	$0;56\ 6/7^{tr}$	$0;56,54,20^{tr}$	$0;56,54,40^{tr}$
if	*DK*	60	60	60	60	60
then	*KL*	0; 1,12	0; 2,52	0; 3 1/7	0; 3, 5,40	0; 3, 5,20
if	*KL*	$62;10^{tr}$	$62;10^{tr}$	$62;10^{tr}$	$62;10^{tr}$	$62;10^{tr}$
then	*DK*	$3108\ 1/3^{tr}$	1301^{tr}	1187^{tr}	1200^{tr}	1208^{tr}

and

$$DK = \frac{62, 10^{tr}}{1^{tr} - 0;57,8^{tr}} = 1301^{tr}.$$

He also finds the length of the axis of the shadow:

$$KS = \frac{62; 10^{tr}}{1^{tr} - 0;45,52^{tr}} = 264^{tr}.$$

Reinhold makes no comment on his results, but it is obvious from what follows that he found them unsatisfactory. Since the final value accepted by Reinhold is 1208^{tr}, it appears he wished to obtain a value close to Ptolemy's 1210^{tr}.

The solar distance assuming a shadow-moon ratio of 19:7

Reinhold begins a new page of the manuscript with the title "Our analytic explanation in establishing the ratio of the diameters of the moon and the shadow."[27] What follows immediately, however, is a crossed out paragraph in which he makes a last attempt to achieve a suitable solar distance from a given ratio. The value 19:7 is referred to as "nostrum adsumptum". Since Ptolemy's and Copernicus' values of the ratios produced solar distances which were too large, it appears Reinhold tried to correct them by using a slightly larger value of the ratio (see Table 3.4 for a comparison of the ratios). He finds the following:

$$MR = 0;17^{tr} \cdot \frac{19}{7} = 0;46\ \ 1/7^{tr},$$

$$LOP = 2^{tr} - 0;46\ \ 1/7^{tr} = 0;73\ \ 6/7^{tr},$$

$$OP = 0;73\ \ 6/7^{tr} - 0;17^{tr} = 0;56\ \ 6/7^{tr},$$

and

$$DK = \frac{62; 10^{tr}}{1^{tr} - 0;56\ \ 6/7^{tr}} = 1187^{tr}.$$

While Reinhold makes no comment, he must have found this value too small for he crossed out the paragraph, drew a line, and proceeded with his *analysis.*

The "analysis"

Reinhold begins by assuming a value for the solar distance and from this derives a value for the ratio. This may seem contrived, since the ratio of the lunar and shadow diameters has a definite value and it can be derived from the comparison of two lunar eclipses. Reinhold may have felt, however, that Copernicus' derivation of the ratio in chapter 18, on which Reinhold chooses not to comment, was not based on real eclipse observations.

[27] "Nostra demonstratio *analytike* in constituenda ratione dimetientum lunae et umbrae" (Ms.R.54v).

Table 3.4

SHADOW-MOON RADIUS RATIOS

Ptolemy (accurately)	$\frac{122}{47}$	=	2;35,44,40,...
Ptolemy (approximately)	$\frac{13}{5}$	=	2;36
Copernicus (1)	$\frac{79}{30}$	=	2;38
Copernicus (2)	$\frac{403}{150}$	=	2;41,11,59,...
Reinhold (1)	$\frac{19}{7}$	=	2;42,51,25,...
Reinhold (2)	$\frac{122}{45}$	=	2;42,40

Reinhold assumes a greatest solar distance of 1200^{tr}, which, he says, does not differ too much from Ptolemy's value ("ne a Ptolemaei sententia nimis procul discedamus"). The value of the true lunar radius is fixed at $0;17^{tr}$. In order to compute the ratio, one must find the true shadow radius that corresponds to a solar distance of 1200^{tr}. Since

$$DK = \frac{KL}{1 - OP},$$

it follows that

$$OP = \frac{DK - KL}{DK}.$$

Instead of finding

$$OP = \frac{1200^{tr} - 62;10^{tr}}{1200^{tr}} = 0;56,53,30^{tr},$$

Reinhold finds KL under the assumption that DK is 1^{tr}:

$$KL = \frac{62;10^{tr}}{1200^{tr}} \cdot 1^{tr} = 0;3,6,30^{tr}.$$

However, he has made a mistake, because in the manuscript KL is written as $0;3,5,40^{tr}$. This accounts in part for the solar distance of 1208^{tr}, rather than 1200^{tr}, which Reinhold obtains when he works the analysis backwards (his following *synthesis*). In triangles KCE and KCD

$$\frac{1}{OP} = \frac{DK}{DK - KL}.$$

Thus if DK is 1^{tr},

$$OP = 1 - KL = 1^{tr} - 0;3,5,40^{tr} = 0;56,54,20^{tr}.$$

The true shadow radius is then

$$\begin{aligned} MR &= 2 - (LO + OP) \\ &= 2^{tr} - 1;13,54,20^{tr} = 0;46,5,40^{tr}. \end{aligned}$$

Reinhold reduces the ratio as follows:

$$\frac{LO}{MR} = \frac{0;17}{0;46,5,40} = \frac{61200}{165940} = \frac{30600}{82970} = \frac{3060}{8297}.$$

At this point he "throws out" one from the 8297 ("His abrice unitatem") and proceeds to find

$$\frac{3060}{8296} = \frac{1530}{4148} = \frac{765}{2074} = \frac{45}{122}.$$

The small error introduced by changing 8297 to 8296 also helps to account for the difference between 1208^{tr} and 1200^{tr}. In the margin opposite his result

Reinhold has written "ratio of the shadow and moon in maximum eccentricity." This is the first time he has specified the eccentricity, and the remark may have been added after his subsequent work with maximum and minimum eccentricity.

The "synthesis"

At this point Reinhold draws another line and begins a new section entitled *synthesis*. The small difference between analysis and synthesis is immediately apparent in the calculation of MR. The ratio 45:122 was obtained from a value of $0;46,5,40^{tr}$ for MR. Working backwards, however,

$$MR = 0;17^{tr} \cdot \frac{122}{45} = 0;46,5,20^{tr}.$$

From this it follows that

$$LOP = 2^{tr} - 0;46,5,20^{tr} = 1;13,54,40^{tr}$$

and

$$OP = 1;13,54,40^{tr} - 0;17^{tr} = 0;56,54,40^{tr}.$$

The solar distance is thus

$$DK = \frac{62:10^{tr}}{1^{tr} - 0;56,54,40^{tr}} = 1208^{tr}.$$

Reinhold concludes this section by finding the solar radius at the distance 1208^{tr}:

$$DC = \frac{1208^{tr}}{62;10^{tr}} \cdot 0;17^{tr} = 5;30,20,16^{tr}.$$

Reinhold has written $5;30^{tr}$ "approximately" in the text, and the more exact $5;30,20^{tr}$ in the margin. He uses both values in his subsequent commentary.

The solar distance for maximum and minimum eccentricity of the sun in apogee and perigee

There follows in the maauscript what is obviously a new section, though it is not so indicated. It is accompanied by the diagram in Figure 3.3. Copernicus, firmly believing in the observations of his predecessors, came to the conclusion that the solar eccentricity and the direction of the apsidal line changed irregularly with time. In his solar model, if the sun is at B (see Figure 3.4), the center of the earth's orbit moves on circle AGD in a period of 3434 years, while the center of the circle F moves slowly in the opposite direction in a period of about 53 000 years (*De Revolutionibus* iii.20). The minimum, mean and maximum eccentricities are represented by BD, BF and BA. It is this small circle AGD which Reinhold has reproduced in his diagram. The eccentric radius to

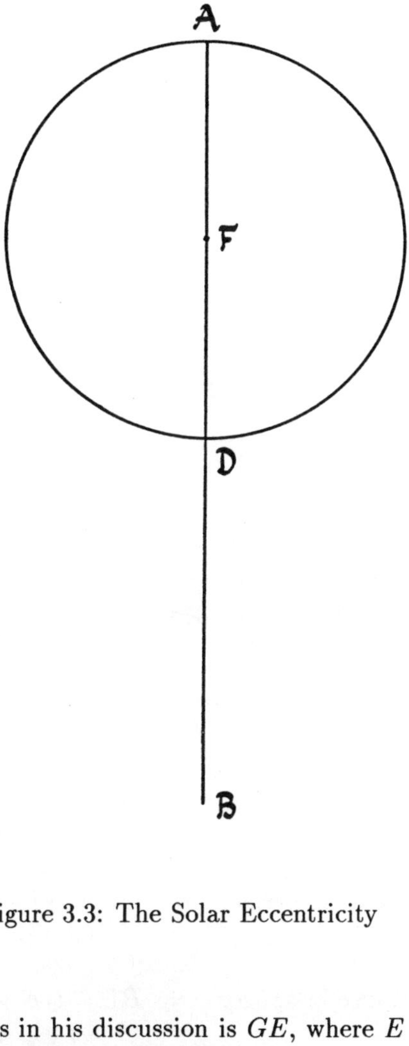

Figure 3.3: The Solar Eccentricity

which Reinhold refers in his discussion is GE, where E is the earth, G is the center of its orbit, and CD equals GE.

In his commentary on *De Revolutionibus* iii,21, Reinhold determined that when the radius of the solar eccentric (GE) is 10 000 000, the maximum solar eccentricity (BA) is 417 000, the mean eccentricity (BF) is 369 456, and the minimum (BD) 321 913. The radius of the small circle ("circellus") is 47 543. Note, that while 47 543 is the difference between BF and BD, it is not the difference between BA and BF. Reinhold restates these values at the beginning of this section, along with the following values for the greatest and least solar distances in apogee and perigee (see Figure 3.5):

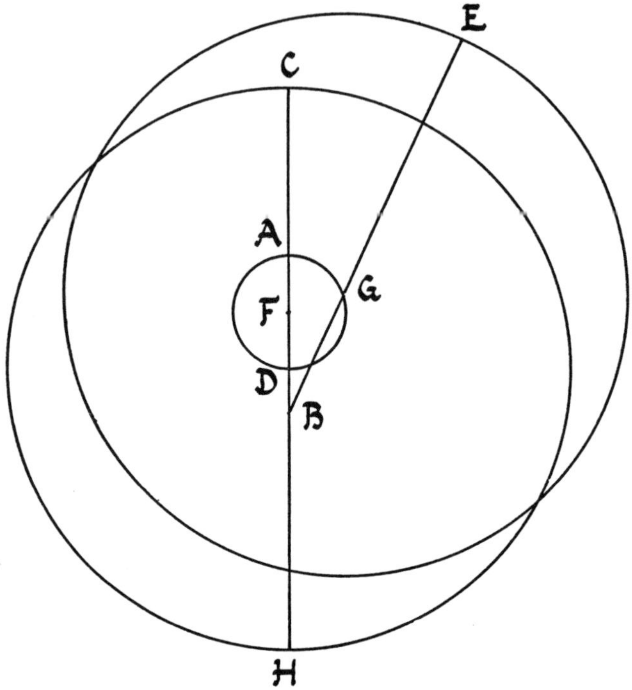

Figure 3.4: Copernicus' Solar Model for the Changing Eccentrity and Apsidal Line

$$
\begin{aligned}
\text{greatest distance in apogee} &= BE = EG + BA \\
&= 10\ 000\ 000 + 417\ 000 \\
&= 10\ 417\ 000, \\
\text{least distance in perigee} &= BE = GE - BA \\
&= 10\ 000\ 000 \text{ - } 417\ 000 \\
&= 9\ 583\ 000, \\
\text{least distance in apogee} &= BE = GE + BD \\
&= 10\ 000\ 000 + 321\ 913 \\
&= 10\ 321\ 913, \\
\text{and} \quad \text{greatest distance in perigee} &= BE = GE - BD \\
&= 10\ 000\ 000 + 321\ 913 \\
&= 9\ 678\ 087.
\end{aligned}
$$

It is at this point that Reinhold remarks that the solar eccentricity was at its maximum in the time of Hipparchus and Ptolemy and is now at a minimum.[28]

Reinhold then converts these relative distances, with unit eccentric radius, to true distances in terrestrial radii (see Table 3.5). Since he has determined

[28] "Fuit autem Hipparchi et Ptolemaei tempore maxima solis eccentrotes, ut nunc est minima" (Ms.R.55r).

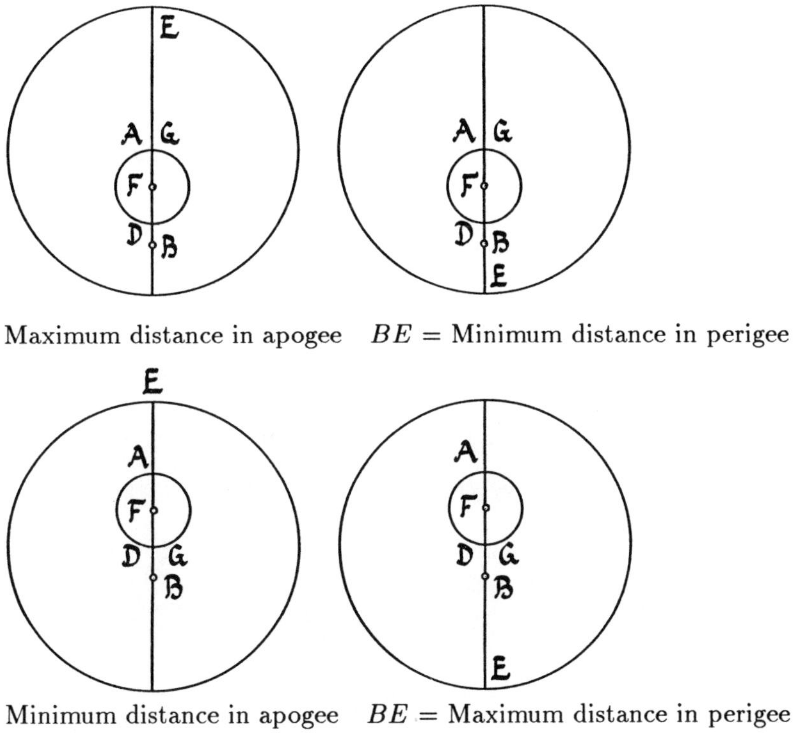

Figure 3.5: Maximum and Minimum Solar Distances in Apogee and Perigee

that the greatest solar distance in apogee is 1208^{tr}, the maximum eccentricity is

$$BA = 417\ 000 \cdot \frac{1208^{tr}}{10\ 417\ 000} = 48;21^{tr}.$$

The mean solar distance is then

$$GE = BE - BA = 1208^{tr} - 48;21^{tr} = 1159;39^{tr},$$

and the least distance in perigee is

$$BE = GE - BA = 1159;39^{tr} - 48;21^{tr} = 1111;18^{tr}.$$

Reinhold's values for the least distance in apogee and the greatest distance in perigee are not quite correct, nor do they agree with his value for the mean distance. The minimum eccentricity is computed correctly:

$$BD = 321\ 913 \cdot \frac{1208^{tr}}{10\ 417\ 000} = 37;19,49^{tr},$$

for which Reinhold has 37;20^{tr}. This implies that the least distance in apogee is

$$BE = GE + BD = 1159;39^{tr} + 37;20^{tr} = 1196;59^{tr},$$

and the greatest distance in perigee is

$$BE = GE - BD = 1159;39^{tr} - 37;20^{tr} = 1122;19^{tr}.$$

Computed independently,

$$GE + BD = 10\ 321\ 913 \cdot \frac{1208^{tr}}{10\ 417\ 000} = 1196;58,24^{tr},$$

and

$$GE - BD = 9\ 678\ 087 \cdot \frac{1208^{tr}}{10\ 417\ 000} = 1122;18,45^{tr}.$$

Reinhold has 1197;0^{tr} for the least distance in apogee, and 1122;20^{tr} for the greatest distance in perigee. These numbers agree among themselves, i.e.,

$$GE + BD - 2BD = 1197;0^{tr} - 74;40^{tr} = 1122;20^{tr}.$$

They do not agree with the mean distance, however, and indeed, Reinhold has added the following note in the margin:

$$\begin{array}{c} 1197{:}\ 0 \\ \underline{1122;20} \\ \underline{2319;20} \end{array}$$

Therefore the mean distance
1159;40.

It is a small discrepancy. Finally, Reinhold finds the diameter of the small circle

$$AD = BA - BD = 48;21^{tr} - 37;20^{tr} = 11;1^{tr}.$$

The angular radius of the sun and the lunar distance at which the solar and lunar diameters appear equal for minimum eccentricity of the sun in apogee

Reinhold found (section 6 above) that the solar radius for greatest distance of the sun in apogee was 5;30,20^{tr}. Using only his approximate value (5;30^{tr}), if the least solar distance in apogee, 1197;0^{tr}, is taken as 10 000 000, then

$$\begin{aligned} DC' &= DC \cdot \frac{10\ 000\ 000}{DK} \\ &= 5;30^{tr} \cdot \frac{10\ 000\ 000}{1197^{tr}} = 459\ 482. \end{aligned}$$

Thus

$$\tan \angle DKC = 0.00459482$$

Table 3.5

REINHOLD'S GREATEST AND LEAST SOLAR DISTANCES IN APOGEE AND PERIGEE

	Distance in parts	Distance in Terrestrial Radii	Solar Radius	Lunar distance for equal radii	Shadow axis	Shadow-moon ratio	Shadow radius
Greatest solar distance in apogee (maximum eccentrity)	10 417 000	1208	$5;30,20^{tr}$	$62;10^{tr}$	268^{tr}	122:45	$0;46,\ 5,20^{tr}$
Least solar distance in apogee (minimum eccentrity)	10 321 913	1197;0	0;15,48	$61;40^{tr}$	266^{tr}	122;45	$0;46,\ 5,25^{tr}$
Greatest solar distance in perigee (minimum eccentrity)	9 678 087	1122;20			249^{tr}		
Least solar distance in perigee (maximum eccentrity)	9 583 000	1111;18			247^{tr}		
Mean solar distance	10 000 000	1159;39					
Maximum eccentrity	417 000	48;21					
Mean eccentricity	369 456						
Minimum eccentrity	321 913	37;20					
Small circle radius	47 543	11; 1					

and

$$\angle DKC = 0;15,48^o.$$

Then, in similar triangles KLO and KDC, if DC is 5;30^{tr}, LO is 0;17^{tr} and DK is 1197^{tr}, the lunar distance is

$$KL = 1197^{tr} \cdot \frac{0;17^{tr}}{5;30^{tr}} = 61;39,49^{tr}.$$

This could also have been found as

$$KL = 0;17^{tr} \cot 0;15,48^o = 61;39,49^{tr}.$$

Reinhold takes 61;40^{tr} for the lunar distance at which the moon has an angular radius of 0;15,48^o.

The length of the axis of the shadow for minimum and maximum eccentricity in apogee and perigee

Reinhold begins a new section entitled "The Axis of the Shadow." It is divided into two parts, the first of which is for the sun at apogee. In Reinhold's final computation of the greatest solar distance in apogee he found a true shadow radius, MR, of 0;46,5,20^{tr}. In similar triangles KES and MRS

$$\frac{KS}{MS} = \frac{KE}{MR},$$

and thus *separando*

$$\frac{KS - MS}{MS} = \frac{KM}{MS} = \frac{KE - MR}{MR}.$$

KM, for greatest solar distance in apogee, is 62;10^{tr}, and KE is 1^{tr}. Then

$$\frac{62;10^{tr}}{MS} = \frac{1^{tr} - 0;46,5,20^{tr}}{0;46,5,20^{tr}} = \frac{0;13,54,40^{tr}}{0;46,5,20^{tr}}.$$

Beneath his values of $KE - MR$ and MR Reinhold has written their equivalents in parts (where 1^{tr} equals 216 000), indicating that he does not perform his computations sexagesimally. Thus he has

$$MS = 62;10^{tr} \cdot \frac{165\ 920}{50\ 080}.$$

The value of MS is then 205;57,46^{tr}, and

$$KS = KM + MS = 62;10^{tr} + 205;57,46^{tr} = 268;7,46^{tr}.$$

Reinhold has written "approximately 268^{tr} " for the axis of the shadow, and beneath this he has written the more exact 268;8^{tr}.

In the previous section, Reinhold computed the lunar distance at which the solar and lunar diameters appear equal for the sun at its least distance in apogee (61;40^{tr}). However, he does not assume that the shadow-moon ratio of 122:45 (for the greatest solar distance in apogee) applies to this case (this is verified in the following section). Thus he must use a different approach to find the length of the shadow axis for least solar distance in apogee. In similar triangles DCS and KES,

$$\frac{DS}{KS} = \frac{DC}{KE},$$

and *separando*

$$\frac{DS - KS}{KS} = \frac{DK}{KS} = \frac{DC - KE}{KE}.$$

In this case, the solar distance DK is 1197^{tr} where the radius of the earth KE is 1^{tr}. Reinhold first computes the length of the shadow axis with his approximate value of the true solar radius, 5;30^{tr},

$$KS = 1197^{tr} \cdot \frac{1^{tr}}{5;30^{tr} - 1^{tr}} = 266;0^{tr},$$

and then uses his more accurate value to obtain

$$KS = 1197^{tr} \cdot \frac{1^{tr}}{5;30,20^{tr} - 1^{tr}} = 265;40^{tr}.$$

This last value is written in the margin.

In the second part of this section, Reinhold finds the length of the shadow axis for the sun in perigee. In the first case, for maximum eccentricity,

$$KS = 1111;20^{tr} \cdot \frac{1^{tr}}{5;30^{tr} - 1^{tr}} = 246;58^{tr},$$

and in the second case, for minimum eccentricity,

$$KS = 1122;20^{tr} \cdot \frac{1^{tr}}{5;30^{tr} - 1^{tr}} = 249;24^{tr}.$$

Reinhold's values are 247^{tr} and 249^{tr}. To the nearest unit, the same results would correspond to a solar radius of 5;30,20^{tr}, but Reinhold has indicated that he uses only 5;30^{tr}.

The shadow-moon ratio for minimum eccentricity of the sun in apogee

In this final section, Reinhold finds the radius of the shadow for least solar distance in apogee. In this case

$$DK = 1197^{tr},$$

$$KS = 266^{tr},$$

$$KL = KM = 61;40^{tr},$$

and

$$SM = KS - KM = 204;20^{tr}.$$

In similar triangles KES and MRS,

$$\frac{MR}{SM} = \frac{KE}{KS},$$

and

$$MR = 204;20^{tr} \cdot \frac{1^{tr}}{266^{tr}} = 0;46,5,24,48^{tr}.$$

Reinhold's value is 0;46,5,25^{tr}, and in the margin he writes that this differs little from the value 0;46,5,20^{tr} for minimum eccentricity. Converting 1^{tr} into 216 000, he proceeds to find the ratio of shadow radius to lunar radius:

$$\frac{MR}{LO} = \frac{0;46,5,25^{tr}}{0;17,0^{tr}} = \frac{165\,925}{61\,200}.$$

This he says is approximately equal to 122:45, the ratio for maximum eccentricity. In fact,

$$\frac{165\,925}{61\,200} = 2;42,40,17,38,...$$

and

$$\frac{122}{45} = 2;42,40.$$

In most sections of his commentary on *De Revolutionibus*, Reinhold's practice is to begin with the assumptions of Copernicus, such as the observational data and the model, and carefully recompute the results. There is no pretense of this in Reinhold's determination of the solar distance. It appears to me that Reinhold had more confidence in Ptolemy's determination of the solar distance than in that of Copernicus.

Reinhold was presented with the following information. Ptolemy found the apparent solar and lunar diameters equal to 0;31,40° when the moon was at its greatest distance in syzygy, 64;10^{tr}. From this, assuming a moon-shadow ratio of 5:13, he derived a solar distance of 1210^{tr}, which Reinhold assumed was the distance of the sun at apogee. In particular, he assumes 1210^{tr} is the greatest solar distance in apogee, in accordance with the Copernican solar model in which the solar eccentricity was near its maximum in Ptolemy's time. Copernicus states that the apparent diameter of the sun in apogee was 0;31,40°, and the angular diameter of the moon at its greatest distance in syzygy (65;30^{tr}) was 0;30°. Assuming a moon-shadow ratio of 150:403, he derived a solar distance of 1179^{tr}. Since the solar eccentricity was near its minimum in Copernicus' time, this should correspond to the least solar distance in apogee.

Reinhold combines Ptolemy's information about the sun with Copernicus' information about the moon to obtain a solar distance in Ptolemy's time. He accepts Copernicus' value of 0;30° for the apparent diameter of the moon at its greatest distance in syzygy and uses his own recomputed value for this distance, based on the Copernican lunar model. He accepts Ptolemy's value of 0;31,20° for the apparent diameter of the sun and derives the greatest solar distance in apogee. The one variable which he does not specify is the apparent diameter of the shadow. Given his assumptions, this is determined by his choice of a moon-shadow ratio. When Reinhold tries different values of the moon-shadow ratio in his essentially trial-and-error attempt to determine the solar distance, the only quantity which actually varies is the shadow diameter.

Reinhold's aim, quite obviously, is to obtain a solar distance in agreement with Ptolemy, rather than Copernicus. The value he finally accepts is 1208^{tr}, compared to Ptolemy's value of 1210^{tr}.

It is surprising to me that Reinhold did not notice the error in his analysis which produced 1208^{tr}, rather than 1200^{tr} which he assumed. Reinhold's value for the least solar distance in apogee, 1197^{tr}, is considerably larger than Copernicus' value of 1179^{tr}.

CHAPTER FOUR

THE RELATIVE MAGNITUDES OF THE SUN, MOON, AND EARTH

Immediately following their determinations of the solar distances, both Ptolemy and Copernicus calculate the relative volumes and diameters of the sun, moon, and earth. While the volumes are not essential to their astronomy, they follow as a natural consequence to the diameters. Given the solar and lunar distances at which the sun and moon appear equal, the solar diameter can be found by taking

$$D_s = D_m \cdot \frac{S_s}{S_m} = LO \cdot \frac{DK}{KL}.$$

The relative volumes of the sun, moon and earth are then proportional to the cubes of their diameters. Ptolemy and Copernicus find the solar diameter relative to the earth and moon, the lunar volume relative to the earth's volume and the solar volume relative to the volumes of both the earth and the moon.

1. PTOLEMY

Ptolemy found in *Almagest* v.15 and 16 that for the new or full moon at apogee, the lunar distance was $64;10^{tr}$, the lunar radius was $0;17,33^{tr}$, and the solar distance was 1210^{tr}. Therefore the solar radius is

$$DC = 0;17,33^{tr} \cdot \frac{1210^{tr}}{64;10^{tr}} = 5;30,56^{tr}.$$

Ptolemy has $5;30^{tr}$. If the radius of the moon is 1, the radius of the earth is

$$KE = 1 \cdot \frac{1^{tr}}{0;17,33^{tr}} = 3;25,7,$$

and the radius of the sun is

$$DC = 5;30 \cdot \frac{1^{tr}}{0;17,33^{tr}} = 18;48,12,$$

or

$$DC = 5;30 \cdot 3;25,7 = 18;48,8.$$

Ptolemy states that the earth's diameter is nearly 3 2/5 times that of the moon, and the sun's 18 4/5.

Accurately computed,

$$1^3 = 1,$$

$$(3\ 2/5)^3 = 39;18,14,$$

and

$$(18\ 4/5)^3 = 6644;40,19.$$

Ptolemy has 39 1/4 for the cube of 3 2/5, and 6644 1/2 for the cube of 18 4/5. Therefore, if the volume of the moon is one, the volume of the earth is 39 1/4, and the volume of the sun is 6644 1/2.

Finally

$$\frac{V_s}{V_e} = \frac{V_s}{V_m} \cdot \frac{V_m}{V_e} = \frac{6644\ 1/2}{39\ 1/4} = 169;17,11.$$

Ptolemy has 170 as the ratio of the solar volume to that of the earth.

2. REINHOLD'S RECOMPUTATION OF PTOLEMY'S MAGNITUDES OF THE SUN, MOON, AND EARTH

In his recomputation, Reinhold, like Ptolemy in the *Almagest*, assumes only one solar distance and assumes that the lunar and solar diameters are equal when the moon is at the apogee of its epicycle in sygyzy. In order to avoid sexagesimal arithmetic, he has converted his own values of LO, KL and DK as follows (where 1^{tr} equals 3600):

$$LO = 0;17,35^{tr} = 1\ 055,$$

$$KL = 64;19,49^{tr} = 231\ 589,$$

and

$$DK = 1193;45,28^{tr} = 4\ 297\ 528.$$

The solar radius is then

$$DC = 1055 \cdot \frac{4\ 297\ 528}{231\ 589} = 19\ 577.$$

Reinhold says that this is "approximately" $5\ 4/9^{tr}$.[1] If the radius of the moon is 1, the radius of the earth is

$$KE = \frac{3600}{1055} = 3;24,44.$$

Reinhold has "approximately" 3 2/5, or 3;24.

Before computing the relative volumes of the sun, moon and earth, Reinhold cites two propositions from Euclid, by book and proposition number: xii.18,

[1] 5 4/9 is 5;26,40, and 19 577 converts into 5;26,17. Reinhold has written in the margin that Ptolemy had $5\ 1/2^{tr}$.

"Spheres are to one another in the triplicate ratio of their respective diameters," and xi.33, "Similar parallelepipedal solids are to one another in the triplicate ratio of their corresponding sides." In Reinhold's time, "diameters" was taken literally as referring to a line segment, not a number which could simply be multiplied, or in this case, cubed. Thus the need for the second proposition.

In Reinhold's commentaries on both Ptolemy and Copernicus, his actual computations rarely appear. The calculation in the commentary on the *Almagest* of the cubes of the diameters of the sun, moon and earth, and the subsequent division to find the ratios of their volumes, is an exception. The results of the multiplications are

$$1055^3 = 1\ 174\ 241\ 375,$$

$$3600^3 = 46\ 656\ 000\ 000,$$

and

$$19\ 577^3 = 7\ 503\ 060\ 053\ 033.$$

Reinhold does not abbreviate these numbers. He divides 7 503 060 053 033 by 1 174 241 375, using the galley method of division, common in the sixteenth century. The division in the manuscript is his actual computation, as can be seen from his practice of scratching out the numbers which have been used.[2] His results are

$$\frac{7\ 503\ 060\ 053\ 033}{1\ 174\ 241\ 375} = 6\ 389,$$

which he says is 6 390 "approximately,"[3]

$$\frac{7\ 503\ 060\ 053\ 033}{46\ 656\ 000\ 000} = 161,$$

with the qualification "approximate,"[4] and

$$\frac{46\ 656\ 000\ 000}{1\ 174\ 241\ 375} = 39\ 3/4,$$

again with the same qualification.[5] Reinhold's conclusion is that the body of the sun contains the body of the moon 6 389 times, but the earth 161 times, while the body of the earth contains that of the lunar body 39 3/4 times.

3. REINHOLD'S EPILOGUE TO PTOLEMY

This concludes Reinhold's mathematical analysis of chapters 12 through 16 of book five of the *Almagest*. It is a self-contained unit, as is emphasized by Reinhold's epilogue at this point. The epilogue is divided into an ***analysis*** and *synthesis*, and I quote it in full as an example of one of the few discursive passages in Reinhold's work.

[2] For a description of the galley or scratch method of division, see Boyer [1], pp. 239-241.

[3] Accurately it is 6 389.7.

[4] Accurately it is 160.8.

[5] Accurately it is 39.741.

Epilogos Totius Huius Methodi

I. Per *analysin* seu regressum ab ultimo, quod nobis proximum est, ad primum illud, quod natura ipsa primum offert seu monstrat.

1. Conditurus erat Ptolemaeus canonem parallaxium solis et lunae ad supputandos defectus solares quae pro diversitate climatum variant. Hic scopus est Ptolemaei.

2. Iam ad condendum hunc canonem parallaxium lunae, opus erat praenosse quae sit proportio distantiae lunaris a centro terrae et semidiameter eiusdem terrae, hoc est, quanta sit lunaris distantia semidiametris terrae mensurata. Similiter et in sole, oportebat praenosse eandem proportionem quae erat inter distantiam solarem et semidiametrum terrae.

3. Verum hanc, seu distantiam solarem seu proportionem illius et semidiametri terrenae, non poterat Ptolemaeus venari ex parallaxi, ut in luna, cum sensu nulla parallaxis solis percipiatur. Ideo alia via instituenda erat ut eam distantiam ratiocinaretur ex distantia lunari et diametro eiusdem lunae apparenti, quae par est diametro solari secundum sensum.

4. Diameter autem lunae apparens exacte saltem sciri non poterat absque collatione duarum eclipsium inaequalium, ut in quarum altera dimidiata pars, in altera vero quadrans solum diametri lunae fuisset obscurata. Huius autem rei investigatio simul ostendit proportionem diametrorum corporis lunae et umbrae iuxta transitum seu distantiam lunae.

5. Distantia vero lunae a terris, seu proportio illius distantiae et semidiametri terrae, colligi potest ex observatione in qua per regulas parallaticas Ptolemaei, vel similia instrumenti, cognita sit distantia lunae apparens a vertice, quae a vera distantia euisdem lunae a vertice non parum discrepat. Ex utriusque igitur collatione seu parallaxi, erudite geometer noticiam distantiae [sic] lunae a terris eruere potest.

6. Ut autem per numerationem vera distantia lunae a vertice cognosci queat, tenere oportet noticiam verae latitudinis lunae.

7. Vera denique latitudo lunae colligi poterat ex tali observatione in qua luna quam proxime ad verticem capitis accederet, ut Alexandriae existens circa principium Cancri seu *tropen* aestivam, et distaret a nodis utrimque circiter quadrantem circuli. Ita enim luna expers erat parallaxeos et versabatur in maxima latitudine. Qua quidem data, iuxta doctrinam sphaericorum triangulorum facile est canonem verae latitudinis lunae componere.

Ita tandem *analysis* monstrat et viam, et quid primo omnium inquirendum est, ut caetera ad supputationem canonis parallaxeon necessaria cognoscere possimus.

II. *Synthesis* quae orditur ab eo quod ultimum monstravit *analysis* tanquam principium, unde reliqua investigatio proficisci debeat.

1. Primum ergo iuxta maximam latitudinem lunae condendus fuit canon latitudinis lunae verae.

2. Ex hoc canone semper dato vero loco lunae, vera eius distantia a vertice numerari potest.

3. Ad hanc veram distantiam lunae a vertice collata apparens eiusdem lunae distantia a vertice, exhibet *parallaxin* lunae.

4. Hinc geometer iuxta doctrinam planorum triangulorum ratiocinatur distantiam lunae a terris, seu proportionem lunaris distantiae et semidiametri terrae.

5. Inde ex collatione duarum eclipsium lunae, et apparentem eius diametrum et proportionem semidiametrorum lunae et umbrae scrutatur.

6. Hinc pergit ad investigationem distantiae solaris, seu proportionis quae est inter distantiam solis a terra et semidiametri terrae.

Cum hoc autem simul tanquam *porismata* demonstrantur haec:

1. Quod umbra terrae *konoeides,*
2. Quod sol maior terra,
3. Quod luna minor terra,
4. Et quae sit horum 3 corporum proportio quod ad diametros eorum adtinet.

7. His omnibus ita praeparatis, patet nunc aditus ad condendum canonem parallaxium solis et lunae. (Ms.R.184v-186v)

Epilogue of this Whole Method

I. Through an *analysis*, or a regression from the end, which is closest to us; to the beginning, which nature itself offers or shows in the first place.

1. Ptolemy had constructed the parallax table of the sun and moon for the computation of solar eclipses, which vary with different latitudes. This is the goal of Ptolemy.

2. Now for the construction of the parallax table of the moon, it was necessary to determine beforehand the ratio of the distance of the moon from the center of the earth and the radius of the same earth, i.e., how much the lunar distance is measured in earth radii. Similarly for the sun, it was necessary to know the same proportion which was between the solar distance and the radius of the earth.

3. Ptolemy had not been able to find from the parallax [of the sun] either the solar distance or its proportion to the radius of the earth, as in the case of the moon, since the solar parallax is not perceptible to the senses. Therefore another way was established so that this distance could be calculated from the lunar distance and the apparent diameter

of the same moon, which is equal to the solar diameter according to the senses.

4. However, the apparent diameter of the moon could not be known exactly without comparing two unequal eclipses, such that in one a half, in the other only a quarter, of the lunar diameter was obscured. The investigation of this problem also showed the proportion of the diameters of the body of the moon and the shadow at the place of transit or the distance of the moon.

5. The distance of the moon from the earth, or the proportion of this distance and the radius of the earth, can be computed from observations in which, by means of the parallactic rulers of Ptolemy, or similar instruments, the apparent distance of the moon from the zenith is known, which differs significantly from the true distance of the same moon from the zenith. Therefore, from the comparison or parallax of both [zenith distances], the geometer can cleverly find the distance of the moon from the earth.

6. In order to know by calculation the true distance of the moon from the zenith, it is necessary to know the true latitude of the moon.

7. Finally, he was able to compute the true latitude of the moon from an observation in which the moon approached as close as possible to the zenith, which in Alexandria is at about the beginning of Cancer or at the summer tropic, and [in which the moon] is about a quarter of a circle distant from the nodes on either side. For then the moon was free from parallax and at its greatest latitude. Given this, a table of the true lunar latitude is easily constructed from the theory of spherical triangles.

Thus at last *analysis* shows both the method and that which was sought first of all, so that we can determine the remaining things necessary for the computation of a parallax table.

II. A *synthesis* which begins from that which the *analysis* showed last, from which the remaining investigation ought to proceed.

1. First, therefore, a table of the true latitude of the moon was made from the maximum latitude.

2. From this table, always given the true position of the moon, it is possible to calculate its true zenith distance.

3. The apparent zenith distance of the moon, compared to the true zenith distance of the moon, shows the parallax of the moon.

4. With the theory of plane triangles, the geometer then calculates the distance of the moon from the earth, or the proportion of the lunar distance and the radius of the earth.

5. Then by a comparison of two eclipses of the moon, both its apparent diameter and the proportion of the diameters of the moon and shadow are sought.

6. Next one proceeds to an investigation of the solar distance, or the proportion between the distance of the sun from the earth and the radius of the earth.

Together with this, the following are proved as *porisms*:

1. That the shadow of the earth is conical
2. That the sun is larger than the earth
3. That the moon is smaller than the earth
4. and what is the proportion of these three bodies, which is related to their diameters.

7. Thus with all of these things supplied, it is now possible to begin the construction of a table of parallax of the sun and moon.

Though there are seven points in both the analysis and synthesis, the correspondence is not one to one. Point 2 in the analysis is merely a general statement of points 3 and 4, and point 5 of the analysis is divided into steps 3 and 4 in the synthesis. The four porisms in the synthesis summarize Reinhold's previous discussion of why the shadow of the earth is conical.

Reinhold's concluding words are especially important, as they date his commentary on the *Almagest*: "*Theo doxa*, Calendis Januarii, 1549" (Ms.R.186v). Reinhold refers to his work on Ptolemy in the Copernican commentary, and thus the commentary on *De Revolutionibus* must have been written subsequent to January 1, 1549, i.e., not much before the publication of the *Prutenic Tables* in 1551.

4. COPERNICUS

The method of computing the solar radius implies that triangles DCK and LOK are similar, i.e., one must use the solar and lunar distances at which the apparent diameters of these bodies appear equal. In *De Revolutionibus* iv.20, Copernicus does not state the solar and lunar distances which he assumed, but they can be reconstructed from his results.

Copernicus gives the following data:

$$\frac{KL}{DK} = \frac{1}{18},$$

$$\frac{LO}{DC} = \frac{1}{18},$$

and

$$DC = 5;27^{tr}.$$

In *De Revolutionibus* iv.19, Copernicus found that the solar and lunar diameters appear equal when the sun is at its greatest distance in apogee (and implicity near minimum eccentricity), under which circumstances the lunar distance is 62^{tr}. Now

$$\frac{65;30}{1179} = \frac{1}{18}$$

exactly, whereas

$$\frac{62}{1179} = \frac{1}{19;1}.$$

Thus Copernicus has used his value for the greatest distance of the moon in syzygy, just as he first used this distance in his determination of the solar distance. While he corrects this mistake in chapter 19, he does not do so in chapter 20.

Copernicus does not state the radius of the moon until after he has computed the ratio of the volumes of the sun and earth, when he gives 0;17,9^{tr} for LO. The solar radius is

$$DC = LO \cdot \frac{DC}{LO}.$$

Thus one would expect, from his stated value of the lunar radius,

$$DC = 0;17,9^{tr} \cdot 18 = 5;8,42^{tr}.$$

Copernicus has 5;27^{tr}. Now

$$0;18,11^{tr} \cdot 18 = 5;27,18^{tr}.$$

Thus Copernicus has used his initial value of the lunar radius from chapter 19, based on the lunar distance 65;30^{tr} and the shadow-moon ratio 79:30. Copernicus then confirms his value for the solar radius by considering similar triangles DCS and KES, in which

$$\frac{DC}{KE} = \frac{DS}{KS}.$$

Knowing the solar distance and the length of the axis of the cone, this becomes

$$\begin{aligned} DC &= \frac{DS}{KS} \cdot KE \\ &= \frac{1179^{tr} + 265^{tr}}{265^{tr}} \cdot 1^{tr} = \frac{1444^{tr}}{265} = 5;26,56^{tr} \cong 5;27^{tr}. \end{aligned}$$

Copernicus next says that since spheres ("globi") are in the triplicate ratio of their diameters, and since

$$5;27^3 = 161\ 7/8,$$

the sun is 161 7/8 times as great as the earth.[6] He gives the ratio of lunar and terrestrial radii as[7]

$$\frac{1}{0;17,9} = \frac{7}{2}.$$

[6] Accurately, $5;27^3$ is 161;52,43, and 161 7/8 is equal to 161;52,30.

[7] Accurately,

$$\frac{1}{0;17,9} = \frac{7}{2;0,3}.$$

Since

$$\frac{1}{0;18,11} = \frac{7}{2;7,17},$$

it appears that here Copernicus is actually using 0;17,9, and not 0;18,11, as earlier. Taking the cube of this ratio,

$$\frac{7^3}{2^3} = \frac{343}{8} = 42\ 7/8$$

exactly. This is Copernicus' value. Finally, Copernicus gives the ratio of solar and lunar volumes. This can be found simply from

$$\frac{V_s}{V_m} = \frac{V_s}{V_e} \cdot \frac{V_e}{V_m} = 161\ 7/8 \cdot 42\ 7/8 = 6\ 940\ 25/64,$$

or more precisely, using a lunar radius of $0;17,9^{tr}$,

$$\frac{V_s}{V_m} = \frac{5;27^3}{0;17,9^3} = \frac{161;52,30}{0;1,24} = 6\ 937;30.$$

Copernicus has 6 937.[8]

5. REINHOLD'S RECOMPUTATION OF COPERNICUS' MAGNITUDES OF THE SUN, MOON, AND EARTH

In his commentary on *De Revolutionibus* iv.19, Reinhold found that the solar radius was approximately $5;30^{tr}$. Here he states the ratio of solar and terrestrial diameters as

$$\frac{DC}{KE} = \frac{11}{2}.$$

Converting terrestrial radii to parts, where 1^{tr} equals 60 parts, Reinhold uses 60,17, and 330 for the radii of the earth, moon, and sun, respectively, for his comparison of the volumes of the sun, moon and earth. He gives only his results, which are correct to the nearest unit:

$$\frac{V_s}{V_e} = \frac{330^3}{60^3} = \frac{35\ 937\ 000}{216\ 000} = 166.3 \cong 166,$$

$$\frac{V_e}{V_m} = \frac{60^3}{17^3} = \frac{216\ 000}{4\ 913} = 43.9 \cong 44,$$

and

$$\frac{V_s}{V_m} = \frac{330^3}{17^3} = \frac{35\ 937\ 000}{4\ 913} = 7\ 314.6 \cong 7\ 315.$$

[8] Copernicus writes this number as "septies milies minus lxiii." The printed edition has changed this to "septies milies minus lxii."

CHAPTER FIVE

SOLAR AND LUNAR PARALLAXES AND DIAMETERS

The geocentric distances of the sun and moon allow one to compute their parallaxes and diameters. These quantities are essential for the accurate prediction of eclipses. For solar eclipses, one must know the lunar parallax. Eclipse limits are determined from the apparent radii of the sun, moon and shadow. In the *Almagest*, Ptolemy assumes a single solar distance and thus an unvarying solar radius of 0;15,40°. In *Almagest* v.14, he found an apparent lunar radius of 0;15,40° for the moon at apogee, and in *Almagest* vi.5 he computes a radius of 0;17,40° for the moon in perigee. In both cases he finds the shadow radius to be approximately 2 3/5 times the lunar radius. The extremal values are incorporated into Ptolemy's eclipse tables and he does not provide separate tables for the solar, lunar and shadow radii. Copernicus devotes three chapters to the discussion of solar parallax and diameter, lunar parallax and diameter, and the moon-shadow ratio (*De Revolutionibus* iv.21, 22 and 23). In the present chapter I shall analyse this section of *De Revolutionibus*, Reinhold's commentary on these three chapters, and the parallax tables of Ptolemy, Copernicus and Reinhold.

1. THE APPARENT DIAMETERS OF THE SUN AND ITS PARALLAXES

The solar diameter and parallax vary with the distance of the sun from the earth. In chapter 21 Copernicus attempts to find the limits within which they vary. His statements concerning distance and parallax are rather imprecise, and his values for the solar diameter do not agree with the values he assumed in chapter 19.

Implicit in Copernicus' investigation of solar distance and parallax is the assumption of minimum solar eccentricity. Copernicus says that the farthest ("remotissimus") solar distance is 10 322, where the radius of the eccentric is 10 000, and the nearest ("proximus") distance is 9 678. Strictly speaking, these correspond to the least distance in apogee and the greatest distance in perigee. The solar distance in parts is derived in book iii of *De Revolutionibus.* In chapter 18, Copernicus finds a value of 10 322 for his own time (A.D.1515). He subsequently changed this in the manuscript to 10 323, and it is this latter value which occurs in the Nuremberg edition. In chapter 21 of book iv, Copernicus

Table 5.1

COPERNICUS' GREATEST AND LEAST SOLAR DISTANCES

	Distance in parts	Distance in Terrestrial Radii	True solar Radius
Least solar distance in apogee	10 323	1179^{tr}	$5;27^{tr}$
Greatest solar distance in perigee	9 678	1105^{tr}	
Mean solar distance	10 000	1142^{tr}	

again has the original manuscript value of 10 322, which is again changed to 10 323 in the Nuremberg edition. In chapter 21 of book iii, he found a least solar distance in in apogee of 10 321. Thus it appears that in book iii Copernicus distinguishes between the solar distance in his own time and the solar distance at the time of absolute minimum eccentricity. Since the solar distance in apogee derived in book iv is also for his own time, Copernicus is correct in saying that 1179^{tr} correspond to 10 323. Presumably the distinction is not necessary when the solar distance is measured in terrestrial radii. Copernicus now derives the solar distance in perigee in terrestrial radii by multiplying 1179^{tr}, the distance in his own time in terrestrial radii, by the ratio of the distances in parts:

$$S_s = 1179^{tr} \cdot \frac{9\ 678}{10\ 323} = 1105^{tr}.$$

From this it follows that the mean distance in terrestrial radii is

$$\frac{1179^{tr} + 1105^{tr}}{2} = 1142^{tr}$$

(see Table 5.1).

Copernicus next finds the solar parallax when the sun is at apogee and perigee. In Figure 5.1, which is taken from Reinhold's manuscript,[1] A is the center of the earth, B is the observer on the surface of the earth, C is the zenith, and D is the position of the sun on the horizon. The solar parallax is then $\angle BDA$. If the terrestrial radius AB is 1^{tr}, and the solar distance AD is first 1179^{tr}, then , when AD is 1 000 000,

$$AB = 1\ 000\ 000 \cdot \frac{1^{tr}}{1179^{tr}} = 848.$$

---

[1] Ms.R.56v. Copernicus refers only to a right triangle, and specifies that the parallax is horizontal.

From Copernicus' table of chords

$$848 = \frac{1}{2} \text{ crd } 2 \cdot 0;2,54,50^o.$$

If the solar distance is 1105^{tr}, then

$$AB = 1\,000\,000 \cdot \frac{1^{tr}}{1105^{tr}} = 905,$$

and

$$905 = \frac{1}{2} \text{ crd } 2 \cdot 0;3,6,35^o.$$

The values given by Copernicus are 0;2,55° and 0;3,7°.

Copernicus proceeds to find the angular diameters of the sun for the solar distances 1179^{tr} and 1105^{tr}. In Figure 5.2, also taken from Reinhold's manuscript,[2] A is the center of the earth, D is the center of the sun, and DB is the solar radius. In chapter 20, Copernicus found that the true solar radius was $5;27^{tr}$. If the solar distance of 1179^{tr} is taken as 2 000 000, then

$$BD = 2\,000\,000 \cdot \frac{5;27^{tr}}{1179^{tr}} = 9\,245,\,^3$$

and

$$9245 = \frac{1}{2} \text{ crd } 2 \cdot 0;31,47^o.$$

If the solar distance is then 1105^{tr},

$$BD = 2\,000\,000 \cdot \frac{5;27^{tr}}{1105^{tr}} = 9864,$$

and

$$9864 = \frac{1}{2} \text{ crd } 2 \cdot 0;33,55^o.$$

The values given by Copernicus are 0;31,48° for apogee and 0;33,54° for perigee. Copernicus' solar distances are based on the assumption that the apparent diameter of the sun in apogee is only 0;31,40°. The table of solar diameters[4] begins with the value 0;31,40° (for 6° of anomaly), not 0;31,48°.

---

[2] Ms.R.56v. Normally, in computing the size and distance of the sun, $\angle ADB$ is taken as a right angle. Here it is $\angle ABD$ which is right,

[3] Copernicus has written 9210 in his manuscript, but 9245 appears in the printed edition. He has obviously used 9245, since

$$9210 = \frac{1}{2} \text{ crd } 2 \cdot 0;31,39^o.$$

[4] The table actually gives radii, not diameters.

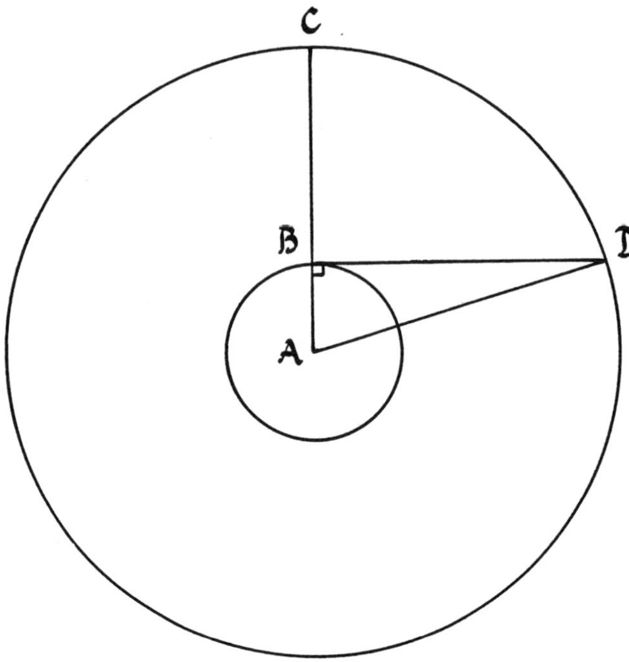

Figure 5.1: The Solar and Lunar Parallax

The apparent diameters at apogee and perigee differ by 0;2,6°,[5] and the parallaxes differ by only 0;0,12°. Copernicus decides that since these differences are so small, he will take the greatest solar parallax to be 0;3° everywhere, i.e., the horizontal parallax of the sun is 0;3° regardless of the solar distance. Ptolemy's value for the horizontal solar parallax was 0;2,51°. Copernicus knew from *Epitome* v.25, however, that al-Battānī adjusted Ptolemy's value to

$$0;2,51^\circ + \frac{1}{19}(0;2,51^\circ) = 0;2,51 + 0;0,9^\circ = 0;3^\circ.$$

1.1 Reinhold's Recomputation (Table 5.2)

In recomputing Copernicus' solar distance, Reinhold found the greatest and least distances in apogee and perigee for both minimum and maximum eccentricity. In computing the apparent diameters of the sun and its parallaxes, he considers the same four cases.

He first finds the solar parallaxes for maximum eccentricity.

[5] Copernicus has originally written 0;2,14° in the manuscript. This value may follow from the use of 9210, rather than 9245, for then

$$0;33,54^\circ - 0;31,39^\circ = 0;2,15^\circ.$$

Table 5.2

COPERNICUS AND REINHOLD – APPARENT DIAMETERS AND PARALLAXES OF THE SUN

	REINHOLD			COPERNICUS	
	Solar Parallax	Apparent Solar Radius $R_s = 5;30^{tr}$	Apparent Solar Radius $R_s=5;30,18^{tr}$	Solar Parallax	Apparent Diameter
Greatest solar distance in apogee (maximum eccentricity)	$0;\ 2,51^{tr}$	$0;15,39^{tr}$	$0;15,40,11^{tr}$		
Least solar distance in apogee (minimum eccentricity)	$0;\ 2,52^{tr}$	$0;15,48^{tr}$	$0;15,48,38^{tr}$	$0;\ 2,55^{tr}$	$0;31,48^{tr}$
Greatest solar distance in perigee (minimum eccentricity)	$0;\ 3,\ 4^{tr}$	$0;16,51^{tr}$	$0;16,51,45^{tr}$	$0;\ 3,\ 7^{tr}$	$0;33,54^{tr}$
Least solar distance in perigee (maximum eccentricity)	$0;\ 3,\ 6^{tr}$	$0;17,\ 1^{tr}$	$0;17,\ 1,46^{tr}$		

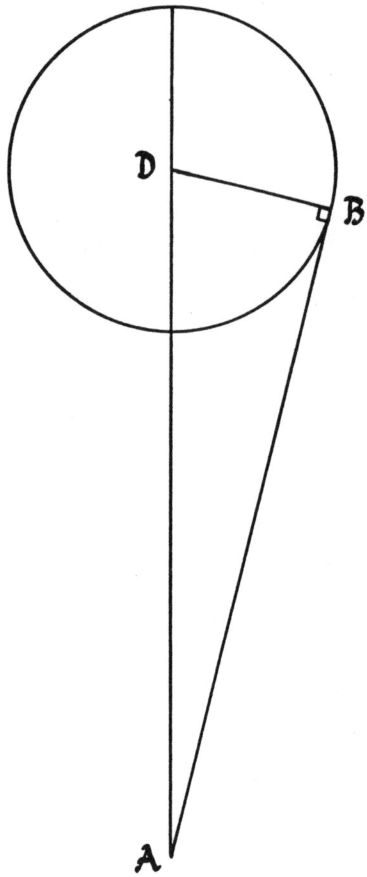

Figure 5.2: The Angular Diameters of the Sun and Moon

For greatest distance in apogee, 1208^{tr} (see Figure 5.1),

$$AB = 100\ 000\ 000 \cdot \frac{1^{tr}}{1208^{tr}} = 82\ 781,$$

and

$$\pi = \sin^{-1}\ 82\ 781 = 0;2,51^{o}.$$

For least distance in perigee, $1111;20^{tr}$,[6]

$$AB = 100\ 000\ 000 \cdot \frac{1^{tr}}{1111;20^{tr}} = 89\ 982,$$

[6] In his commentary on chapter 19, Reinhold found $1111;18^{tr}$ for the least solar distance in perigee. He states only 1111^{tr} in his discussion of solar parallax. Neither of these values will produce his result, which can only be obtained by using $1111;20^{tr}$. Reinhold also states and uses this particular approximation in his calculation of the length of the shadow axis.

and

$$\pi = \sin^{-1}\ 89\ 982 = 0;3,6^o\ .$$

The difference between these two parallaxes is

$$0;3,6^o - 0;2,51^o = 0;0,15^o.$$

Reinhold next computes the parallaxes for minimum eccentricity. At least solar distance in apogee, 1197^{tr},

$$AB = 100\ 000\ 000 \cdot \frac{1^{tr}}{1197^{tr}} = 83\ 542,$$

and

$$\pi = \sin^{-1}\ 83542 = 0;2,52^o.$$

At greatest solar distance in perigee, $1122;20^{tr}$,

$$AB = 100\ 000\ 000 \cdot \frac{1^{tr}}{1122;20^{tr}} = 89\ 100,$$

and

$$\pi = \sin^{-1}\ 89\ 100 = 0;3,4^o.$$

The difference between these two parallaxes is

$$0;3,4^o - 0;2,52^o = 0;0,12^o.$$

In his parallax table, Reinhold uses the mean solar distance of $1159;40^{tr}$, which results in a horizontal parallax of $0;2,58^o$.

Reinhold first computes the apparent solar diameter assuming a true solar radius of $5;30^{tr}$. He has subsequently returned to the manuscript and recomputed his values with a solar radius of $5;30,18^{tr}$. In Reinhold's final computation of the solar distance, he found a solar radius of approximately $5;30^{tr}$, but $5;30,20^{tr}$ exactly. The value $5;30,20^{tr}$ is correct, and I cannot account for $5;30,18^{tr}$.

First, assuming maximum eccentricity, if the greatest solar distance in apogee is 1208^{tr} and the true solar radius is $5;30^{tr}$ (see Figure 5.2),

$$BD = 100\ 000\ 000 \cdot \frac{5;30^{tr}}{1208^{tr}} = 455\ 298$$

and

$$R_s = \sin^{-1}\ 455\ 298 = 0;15,39^o.$$

If the true solar radius is $5;30,18^{tr}$,

$$BD = 455\ 712.$$

Reinhold has 455 804. This value follows precisely from a true solar radius of $5;30,22^{tr}$. The apparent solar diameter is then

$$R_s = \sin^{-1}\ 455\ 804 = 0;15,40,10^o.$$

Reinhold has 0;15,40,11°.

If the least solar distance in perigee is 1111;20^{tr} and the true solar radius is 5;30^{tr},

$$BD = 100\ 000\ 000 \cdot \frac{5;30^{tr}}{1111;20^{tr}} = 494\ 901,$$

and

$$R_s = \sin^{-1}\ 494\ 901 = 0;17,1^o.$$

If the true solar radius is now 5;30,18^{tr},

$$BD = 495\ 351,$$

and

$$R_s = 0;17,1,43^o\ .$$

Reinhold has 0;17,1,46°. The difference between these two solar radii is, in the first case,

$$0;17,1^o - 0;15,39^o = 0;1,22^o,$$

and in the second

$$0;17,1,46^o - 0;15,40,11^o = 0;1,21,35^o.$$

Finally, for minimum eccentricity, if the least solar distance in apogee is 1197^{tr} and the true solar radius is 5;30^{tr},

$$BD = 459\ 482,$$

and

$$R_s = 0;15,48^o.$$

If the true solar radius is 5;30,18^{tr},

$$BD = 459\ 900,$$

and

$$R_s = 0;15,48,37^o.$$

Reinhold has 0;15,48,38°. If the greatest distance in perigee is 1122;20^{tr} and the true solar radius is 5;30^{tr},

$$BD = 490\ 050,$$

and

$$R_s = 0;16,51^o.$$

If the true solar radius is 5;30,18^{tr},

$$BD = 490\ 496,$$

and

$$R_s = 0;16,51,43^o.$$

Reinhold has 0;16,51,45°. In all four cases where Reinhold has increased his value for the true solar radius by one place, he has also found one more sexagesimal place for the apparent solar radius. In each of the four cases, it is the second value which Reinhold has used in his table of solar radii. The differences between the solar radii for minimum eccentricity are

$$0;16,51^o - 0;15,48^o = 0;1,3^o,$$

and

$$0;16,51,45^o - 0;15,48,38^o = 0;1,3,7^o.$$

The various differences in parallax and radii are not actually used. Reinhold computes them simply in imitation of Copernicus.

2. THE APPARENT DIAMETERS OF THE MOON AND ITS PARALLAXES

Copernicus finds the horizontal parallax of the moon for greatest and least distances in syzygy and quadrature (see Table 5.3). In Figure 5.1, where C is the zenith and D is the position of the moon on the horizon,

$$\sin\pi = \frac{AB}{AD} = \frac{1}{S_m},$$

where the radius of the earth is one. Using Copernicus' table of chords and the lunar distances stated in *De Revolutionibus* iv.22, I have found that if the greatest lunar distance in syzygy is 65;30^{tr},

$$\pi_1 = 0;52,30^o,$$

if the least distance in syzygy is 55;8^{tr},

$$\pi_2 = 1;2,22^o,$$

if the greatest distance in quadrature is 68;21^{tr},[7]

$$\pi_3 = 0;50,19^o,$$

and if the least distance in quadrature is 52;17^{tr},[8]

$$\pi_4 = 1;\ 5,46^o.$$

Copernicus' values, which subsequently appear in the parallax table, are

$$\pi_1 = 0;52,24^o,$$

[7] Copernicus has originally written 68;32^{tr} in the manuscript. His value for this distance in chapter 17, where the lunar distance is derived, is 68;20^{tr}.

[8] Copernicus has originally written 52;27^{tr} in the manuscript.

Table 5.3

COPERNICUS' PARALLAXES AND APPARENT DIAMETER5 OF THE MOON

	Lunar Parallax	Apparent Lunar Diameter
Greatest Distance in Syzygy (apogee)	0;52,24°	$\simeq 0;30^o$
Least Distance in Syzygy (perigee)	1; 2,21°	0;35,38°
Greatest Distance in Quadrature (apogee)	0;50,18°	0;28,45°
Least Distance in Quadrature (perigee)	1; 5,45°	0;37,34°

$$\pi_2 = 1;\ 2,21^o,$$

$$\pi_3 = 0;50,18^o,$$

and

$$\pi_4 = 1;\ 5,45^o.$$

Copernicus uses the lunar parallax to find the apparent lunar diameter. He states that the ratio of the parallax to the apparent diameter is equal to the ratio of the radius of the earth to the diameter of the moon. This is a valid approximation, if one assumes that the sines (or chords) of both the parallax and apparent diameter are equal to the angles themselves. In Figure 5.2, if DB is now the lunar radius, and R_m is $\angle BAD$,

$$sin R_m = \frac{DB}{S_m}.$$

Since R_m is small,

$$sin R_m \cong R_m.$$

Then

$$R_m \cong \frac{DB}{S_m} = \frac{DB}{1^{tr}} \cdot \frac{1^{tr}}{S_m}.$$

Now

$$\frac{1^{tr}}{S_m} = sin\pi \cong \pi,$$

since π, the horizontal parallax, is a small angle. Thus

$$R_m \cong \frac{DB}{1^{tr}} \cdot \pi.$$

In chapter 20 Copernicus found

$$\frac{DB}{1^{tr}} = \frac{2}{7}.$$

Thus, in order to find the apparent diameter of the moon, he simply takes

$$D_m = \frac{4}{7}\pi.$$

Copernicus' results are as follows:

$$D_{m1} = 0;30^o,^9$$

$$D_{m2} = 0;35,38^o,$$

$$D_{m3} = 0;28,45^o,$$

and

$$D_{m4} = 0;37,34^o.^{10}$$

The lunar radius column of the diameter table, which is computed for the moon in syzygy, begins with the value 0;15° for apogee, and ends with 0;17,49°, half of 0;35,38°, for perigee. This column is computed with the correct value 0;52,29° for π_1. The value 0;52,24° appears in the parallax table, however. Copernicus mentions, at the end of chapter 22, that the Ptolemaic lunar model would produce a lunar diameter of 1°.

2.1 Reinhold's Recomputation (Table 5.4)

Reinhold's values for the lunar parallaxes are all computed correctly. For a greatest distance of the moon in syzygy of 64;57^{tr}

$$\pi_1 = sin^{-1} 1\ 539\ 646 = 0;52,56^o,$$

for a least distance in syzygy of 54;39^{tr}

$$\pi_2 = sin^{-1} 1\ 829\ 826 = 1;2,54^o,$$

for a greatest distance in quadrature of 67;47^{tr}

$$\pi_3 = sin^{-1} 1\ 475\ 289 - 0;50,43^o,$$

and for a least distance of the moon in quadrature of 51;49^{tr}

$$\pi_4 = sin^{-1} 1\ 929\ 881 - 1;6,21^o.$$

---

[9] Accurately computed, the value is 0;29,57°. Copernicus says "approximately 0;30°. " This value agrees with the apparent diameter for greatest distance in syzygy assumed by Copernicus in *De Revolutionibus* iv.19.

[10] The printed edition has 0;<u>2</u>7,34^{tr}.

Table 5.4

REINHOLD'S PARALLAXES AND APPARENT RADII OF THE MOON

	Lunar Parallax	Apparent Lunar Radius
Greatest Distance in Syzygy	0;52,56°	0;15, 0°
Least Distance in Syzygy	1; 2,54°	0;17,49°
Greatest Distance in Quadrature	0;50,43°	0;14,22°
Least Distance in Quadrature	1; 6,21°	0;18,48°

These four values appear as the horizontal parallax in Reinhold's parallax table (*Prutenic Tables*, f.136r-137r).

Reinhold does not use Copernicus' approximation to compute the angular diameters of the moon. Instead he finds

$$\sin R_m = 100\,000\,000 \cdot \frac{DB}{S_m}.$$

The computations of both the sine values and the angles are done correctly. His results are

$$R_{m1} = \sin^{-1} 436\,233 = 0;15,0^o,$$

$$R_{m2} = \sin^{-1} 518\,451 = 0;17,49^o,$$

$$R_{m3} = \sin^{-1} 417\,999 = 0;14,22^o,$$

and

$$R_{m4} = \sin^{-1} 546\,800 = 0;18,48^o.$$

The first two values, for the moon in syzygy, appear in the column of lunar radii in Reinhold's diameter table (*Prutenic Tables* f.122v).

3. THE SHADOW-MOON RATIO

The object of *De Revolutionibus* iv.23 is to determine the shadow - moon ratio when the sun is at apogee and perigee. Copernicus' treatment of this is somewhat confusing. The shadow-moon ratio varies with both the solar and lunar distances. Since the shadow radius is important only for eclipses, the lunar distances in syzygy are sufficient. Considering only the minimum solar eccentricity, there are four limiting cases of shadow-moon ratios – for greatest and least solar distances when the moon is at greatest distance in syzygy, and for greatest and least solar distances when the moon is at least distance in syzygy. The apparent radius can be found directly given the solar distance, the lunar distance, and the solar radius. The shadow-moon ratios are then the ratios of the individual apparent shadow radii to the respective apparent lunar radii. This is the procedure followed by Reinhold, but not by Copernicus.

In his chapter on the solar distance, Copernicus assumed that the shadow-moon ratio was 403:150. He applied this to the case where the sun was at apogee and the moon was at that distance where solar and lunar diameters appear equal, namely 62^{tr}. In this chapter he finds the shadow radius for the sun at apogee and perigee when the lunar distance is 62^{tr}. This distance has no significance for eclipses, except that when the lunar distance is greater than 62^{tr} a total solar eclipse will be annular.

As in chapter 19, chapter 23 contains in the manuscript two sets of values, the first based on a shadow-moon ratio of 79:30 (for the sun at apogee) and a lunar distance of $65;30^{tr}$, the second based on the ratio 403:150 and a lunar distance of 62^{tr} (see Tables 5.5 and.5.6). The first set of values will only be referred to in footnotes.

In the previous chapter Copernicus found that the apparent lunar diameter for greatest distance in syzygy was $0;30^{tr}$, and for least distance in syzygy, $0;35,38^{tr}$. While he assumed in chapter 19 that the ratio 403:150 was for a lunar distance of 62^{tr}, he uses it here for both $65;30^{tr}$ and $55;8^{tr}$. The ratio is thus associated with the sun in apogee. The true shadow diameters for these two cases are

$$\frac{403}{105} \cdot 0;30^{tr} = 0;80,36^{tr},$$

and

$$\frac{403}{150} \cdot 0;35,38^{tr} = 0;95,44^{tr}.$$[11]

The difference between these two values is

$$0;95,44^{tr} - 0;80,36^{tr} = 0;15,8^{tr}.$$

[11] Using the ratio 79:30, the two shadow diameters are

$$\frac{79}{30} \cdot 0;30^{tr} = 0;79^{tr},$$

and

$$\frac{79}{30} \cdot 0;35,38^{tr} = 0;93,50^{tr}.$$

It is difficult to read Copernicus' original value for this last quantity, since the seconds have been changed to read xliiii. It seems unlikely that he had written 50 seconds, however.

Table 5.5

COPERNICUS' SHADOW-MOON RADIUS RATIO

	Original Ms. Values	Nuremberg edition
Shadow-moon ratio	79;30	403;150
Shadow diameter for greatest lunar distance in syzygy (sun in apogee)	$0;79^{tr}$	$0;80,36^{tr}$
Shadow diameter for least lunar distance in syzygy (sun in apogee)	$0;93,50^{tr}$	$0;95,44^{tr}$
Difference of shadow diameters	$0;14,50^{tr}$	$0;14,8^{tr}$
Solar distance in apogee	1179^{tr}	1179^{tr}
Lunar distance	$65;30^{tr}$	62^{tr}
Shadow radius in terrestrial radii for the above lunar distance	$0;45,10^{tr}$	$0;46,1^{tr}$
Shadow radius in degrees for the above lunar distance	$0;39,30^{\circ}$	$0;42,32^{\circ}$
Length of shadow axis for sun in apogee	265^{tr}	265^{tr}
Solar distance in perigee	1105^{tr}	1105^{tr}
Solar radius minus 1^{tr}	$4;27^{tr}$	$4;27^{tr}$
Shadow axis for sun in perigee	$248;19^{tr}$	$248;19^{tr}$
Solar distance minus lunar distance	$182;49^{tr}$	$186;19^{tr}$
Shadow radius in terrestrial radii for sun in perigee	$0;44,14^{tr}$	$0;45,1^{tr}$
Shadow radius in degrees for sun in perigee	$0;38,41^{\circ}$	$0;41,35^{\circ}$
Difference of shadow radii in terrestrial radi*i**	$0;1,52^{tr}$	$0;1^{tr}$
Difference of shadow radii in degree*s**	$0;1,3\underline{8}^{\circ}$	$0;0,57^{\circ}$
Shadow-moon ratio for sun in apogee	>13:5	>13:5
Shadow-moon ratio for sun in perigee	<13:5	<13:5

* In the manuscript, Copernicus gives the difference of the diameters.

Table 5.6

COPERNICUS' SHADOW RADIUS

	Length of the Shadow Axis	Shadow Radius in Terrestrial Radii		Shadow Radius in Degrees		Shadow-Moon Radius	
		original manuscript	printed edition	original manuscript	printed edition	original manuscript	printed edition
Least solar distance in apogee (minimum eccentricity)	256^{tr}	$0;45,10^{tr}$	$0;46,\ 1^{tr}$	$0;39,30^{o}$	$0;42,32^{o}$	$\frac{0;45,10}{0;17,\ 9} = 3;28$	$\frac{0;46,1}{0;17,9} = 2;41$
Greaest solar distance in perigee (minimum eccentricity)	$248;19^{tr}$	$0;44,14^{tr}$	$0;45,\ 1^{tr}$	$0;38,41^{o}$	$0;41,35^{o}$	$\frac{0;44,14}{0;17,\ 9} = 2;34$	$\frac{0;45,1}{0;17,9} = 2;37$

Copernicus has written $0;14,8^{tr}$.[12]

This determination of the shadow diameters, which is treated in one sentence, has been added in the margin of the manuscript. The chapter originally began by saying that even for a constant lunar distance, the shadow will vary according to the solar distance. Copernicus then proceeds to find the shadow radii for greatest and least solar distances.

Figure 5.3 is reproduced from Copernicus' manuscript. It is similar to the diagram for the solar distance. The line $DKMS$ passes through the sun at D, the earth at K, and the shadow at M. $CERS$ is the tangent line of the shadow cone. In chapter 19 Copernicus found that for a solar distance DK of 1179^{tr} and a lunar distance KM of 62^{tr} , the true shadow radius MR was $0;46,1^{tr}$ and the length of the axis of the shadow was 265^{tr}.[13] He now finds the angular shadow radius. Using the sine to approximate the tangent,

$$\sin \angle MKR \cong \frac{MR}{KM},$$

and

$$\angle MKR = 0;42,32^{o}.$$[14]

This is Copernicus' result for greatest solar distance.

The least solar distance, found in chapter 21, is 1105^{tr}. Copernicus can no longer find the apparent shadow radius by multiplying the lunar radius by the shadow-moon ratio, since this applies to the greatest solar distance. In Figure 5.3, EZ is drawn parallel to DK.
Then in similar triangles CZE and EKS,

$$\frac{CZ}{ZE} = \frac{EK}{KS}.$$

When the terrestrial radius is taken as the unit, the solar radius is $5;27^{tr}$, and the length of the shadow axis is

$$KS = 1^{tr} \cdot \frac{1105^{tr}}{5;27^{tr} - 1^{tr}} = 248;19^{tr}.$$

---

[12] The value $0;15,8^{o}$ does not appear in either the printed edition, or in the manuscript, but is in the Thorn edition, which supposedly was based on the manuscript. For the original set of values, the difference should be

$$0;93,50^{tr} - 0;79^{tr} = 0;14,50^{tr}.$$

This, I think, is the source of the 14. Copernicus has simply forgotten to change it to 15. The minutes in the manuscript seem to read "1" (i.e., 50), which is changed to viii and then to 8.

[13] In agreement with his original set of chapter 19 values, Copernicus has first written here $0;45,10^{tr}$ for the true shadow radius.

[14] Using $0;45,10^{tr}$ for MR and $65;30^{tr}$ for KM, Copernicus originally finds

$$\angle MKR = 0;39,30^{o}.$$

Accurately computed, the value is $0;39,31^{o}$.

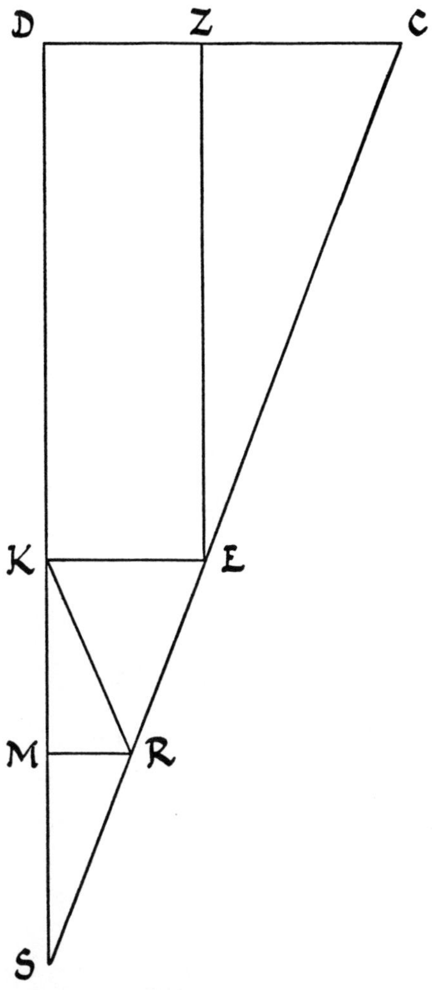

Figure 5.3: The Shadow–Moon Radius Ratio

In similar triangles RMS and EKS,

$$\frac{SM}{MR} = \frac{KS}{KE},$$

and the true shadow radius is

$$MR = 1^{tr} \cdot \frac{248;19^{tr} - 62^{tr}}{248;19^{tr}} = 0;45,1^{tr}.$$[15]

[15] Using a lunar distance of 65;30^{tr},

$$MR = 0;44,10^{tr}.$$

Copernicus has 0;44,14^{tr}.

The angular radius of the shadow is 0;41,36°, for which Copernicus has 0;41,35°.[16] The difference between the true shadow radii[17] for greatest and least solar distance is

$$0;46,1^{tr} - 0;45,1^{tr} = 0;1^{tr},$$

and for the angular radii the difference is

$$0;42,32^{\circ} - 0;41,35^{\circ} = 0;0,57^{\circ}.^{18}$$

Copernicus concludes by saying that for greatest solar distance, the shadow-moon ratio is greater than 13:5 (Ptolemy's value), and for least solar distance, it is less than 13:5. This is based on the assumption of a single value for the lunar distance, 62^{tr}. Thus 13:5 is a mean value of the ratio and one can conveniently assume that the ratio is the same everywhere. The justification of Copernicus' statement about the ratios being greater than and less than 13:5 requires further discussion. For greatest solar distance and a lunar distance of 62^{tr}, the true lunar radius is $0;17,9^{tr}$. Then

$$\frac{0;46,1^{tr}}{0;17,9^{tr}} = 2;41 > 2;36,$$

but

$$\frac{0;45,1^{tr}}{0;17,9^{tr}} = 2;37 > 2;36$$

(where 13/5 equals 2;36). However, using the first set of manuscript values for the shadow radius,

$$\frac{0;45,10^{tr}}{0;17,9^{tr}} = 2;38,$$

and

$$\frac{0;44,14^{tr}}{0;17,9^{tr}} = 2;34.$$

This, I believe, is what Copernicus has done. He first wrote the entire chapter with his original set of values, found that the ratios were greater than and less than 13:5, and did not bother to check these inequalities for his second set of values.

In the shadow radius column of the diameter table, Copernicus does not use 13:5 after all. The manuscript values are computed with 79:30, and the values in the Nuremberg edition with 403:150.

---

[16] With lunar distance $65;30^{tr}$, $\angle MKR$ is 0;38.42°. Copernicus has 0;38,41°.

[17] Copernicus says diameters.

[18] For the original set of values,

$$0;45,10^{tr} - 0;44,14^{tr} = 0;0,56^{tr}$$

and

$$0;39,30^{\circ} - 0;38,41^{\circ} = 0;0,49^{\circ}.$$

Copernicus gives the differences of the diameters.

3.1 Reinhold's Recomputation (Table 5.7)

In his commentary on *De Revolutionibus* iv.19, Reinhold found four values for the length of the shadow axis when the sun was at greatest and least distances in apogee and perigee. He now uses these values to find the true and angular shadow radii for greatest and least lunar distances in syzygy for each of the four solar distances, making eight cases in all. His method for finding the true radius is the same as Copernicus', namely,

$$MR = 1^{tr} \cdot \frac{S_{ax} - S_m}{S_{ax}}.$$

In finding the angular radius he refers to "angulus ex nostro canone," and a comparison of values for sine and tangent reveals that he has solved for the tangent each time:

$$\tan \angle MKR = \frac{MR}{S_m}.$$

His results are summarized in Table 5.7.[19]

Reinhold derived a shadow-moon ratio of 122:45 for greatest solar distance at apogee (maximum eccentricity) in his commentary on the solar distance. After his first two computations, for greatest and least lunar distance in syzygy when the sun is at greatest distance in apogee, he checks his angular shadow radii by multiplying the appropriate angular lunar radii by 122:45. In the first case

$$\frac{122}{45} \cdot 0;15,0^o = 0;40,40^o ,$$

and in the second

$$\frac{122}{45} \cdot 0;17,49^o = 0;48,19^o .$$

After each computation he says that these results do not differ appreciably from his own values, namely 0;40,6° and 0;50,5°.

4. THE PARALLAX AND DIAMETER TABLES

Ptolemy's table of parallax in *Almagest* v.18 lists the parallax in altitude of the sun and moon. The tables found in the *Handy Tables* are much more practical

[19] Reinhold's calculations are relatively free from error. His values for the sine of ∠KMR in the second and third cases are off by one in the last place, but this may be a question of rounding. In Reinhold's verification of the second angular radius, the result should be

$$\frac{122}{45} \cdot 0;17,49^o = 0;48,18,10^o.$$

Reinhold has 0;48,19°. While he states only 0;17,49° for the lunar radius in his commentary on chapter 22, the more accurate value is 0;17,49,22°, and

$$\frac{122}{45} \cdot 0;17,49,22^o = 0;48,19,10^o.$$

Reinhold may have been working with more accuracy than he indicates.

Table 5.7

REINHOLD'S SHADOW RADIUS

	Lenght of the Shadow Axis	Shadow Radius in Terrestrial Radii	Shadow Radius in degrees	$MR = 122/45 \cdot AR_M$
	Greaest solar distance in apogee (1208^{tr})(maximum eccentricity)			
Greatest lunar distance in syzygy	268^{tr}	$0;45,28^{tr}$	$0;40,\ 6^{o}$	$0;40,40^{o}$
Least lunar distance in syzygy	268^{tr}	$0;47,46^{tr}$	$0;50,\ 5^{o}$	$0;48,19^{o}$
	Least solar distance in perigee ($1111;18^{tr}$)(maximum eccentricity)			
Greatest lunar distance in syzygy	247^{tr}	$0;44,13^{tr}$	$0;39,\ 0^{o}$	
Least lunar distance in syzygy	247^{tr}	$0;46,43^{tr}$	$0;48,59^{o}$	
	Least solar distance in apogee (1197^{tr})			
Greatest lunar distance in syzygy	266^{tr}	$0;45,21^{tr}$	$0;40,\ 0^{o}$	
Least lunar distance in syzygy	266^{tr}	$0;47,40^{tr}$	$0;49,58^{o}$	
	Greaest solar distance in perigee ($1122;20^{tr}$)			
Greatest lunar distance in syzygy	249^{tr}	$0;44,21^{tr}$	$0;39,\ 7^{o}$	
Least lunar distance in syzygy	249^{tr}	$0;46,50^{tr}$	$0;49,\ 6^{o}$	

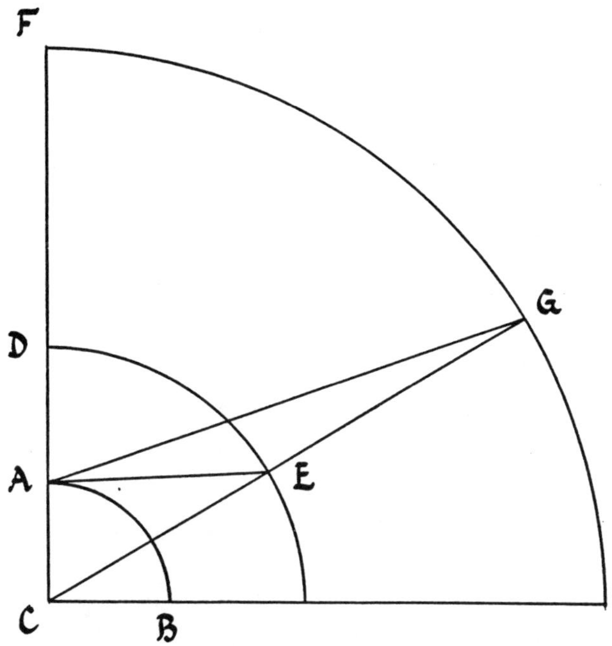

Figure 5.4: Solar and Lunar Parallaxes

and convenient because they display the components in longitude and latitude and the differences in solar and lunar parallax, and these are what one needs for solar eclipses. The parallax table of Copernicus, and subsequently the corresponding table in the *Prutenic Tables*, are for parallax in altitude as in the *Almagest*.

In Figure 5.4, AB is an arc on the surface of the earth with center C, the observer is at A and the zenith at F, the moon is at E, and the sun is at G. The solar parallax in the altitude circle is $\angle AGC$, the lunar parallax is $\angle AEC$, and $\angle GAE$ is the difference between the solar and lunar parallax. The computation of solar parallax as a function of zenith distance, z, is a simple procedure, assuming a single solar distance, S_s. The solar parallax π is found from

$$\sin\pi = \frac{\sin z}{\sqrt{1 + S_s^2 - 2S_s\cos z}} \cong \frac{\sin z}{S_s - \cos z}.$$

The first column of a parallax table contains the argument. Ptolemy's arguments extend from two to ninety degrees, at two degree intervals, Copernicus' arguments extend from 6° to 180° at six degree intervals, and Reinhold's arguments extend from 2° to 180° at two degree intervals. The second column in each table contains the solar parallax as a function of zenith distance. For Ptolemy, this column can be entered directly with z, but for Copernicus and Reinhold the argument must be $2z$. The solar parallax is computed for a mean solar distance: 1210^{tr} for Ptolemy, 1142^{tr} for Copernicus and $1159;40^{tr}$

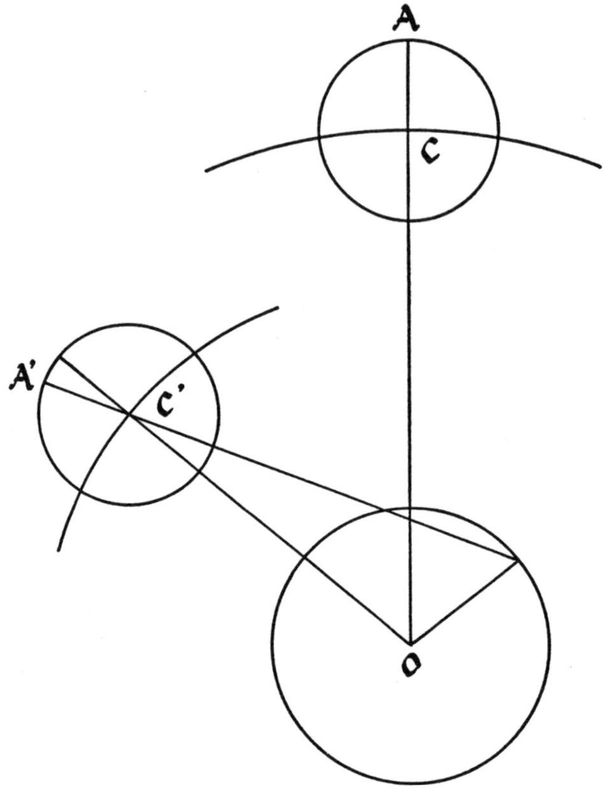

Figure 5.5: The Ptolemaic Lunar Model

for Reinhold. The remaining columns in all three tables concern the lunar parallax.

4.1 Ptolemy's Lunar Parallax

The distance of the moon in Ptolemy's model varies considerably, and thus the lunar parallax must vary accordingly. The method used by Ptolemy in the *Almagest* is first to calculate the lunar parallax for four extreme distances: greatest and least distances in syzygy and quadrature. These values are found in the first four columns of lunar parallax, columns c_3 through c_6, as a function of zenith distance. The lunar parallax at intermediate distances is then found by interpolating between the values for the extreme distances. There are two elements in the Ptolemaic model which determine the lunar distance: the position of the moon on the epicycle and the distance of the center of the epicycle from O (see Figure 5.5). Column 7 of the parallax table is an interpolation function for syzygy, where the epicycle is at the apogee of the deferent and the lunar distance varies as a function of lunar anomaly. Column 8 is a similar function for the case of quadrature, when the epicycle makes its closest approach to the earth. Finally, column 9 provides an adjustment for an intermediate position of the epicycle, as a function of the elongation of the moon from the mean sun.

In Figure 5.6, O is the observer and center of the small circle in Ptolemy's lunar model (with radius e), C is the center of the epicycle at syzygy, and C' is the center of the epicycle at quadrature. The four extreme distances are indicated by the numbers 1 through 4, and the parallaxes at these positions may also be denoted by these four numbers. The lunar parallax at syzygy varies between a minimum of π_1 at distance OA, and a maximum of π_2 at distance OB. In Figure 5.7, the moon at D is in an intermediate position between A and B. The lunar anomaly γ is $\angle ACD$, and r is the epicycle radius. The lunar distance is

$$S_m(\gamma) = \sqrt{(OC + r\cos\gamma)^2 + (r\sin\gamma)^2}.$$

Ptolemy uses this distance to produce a simple linear interpolation between maximum and minimum parallax at syzygy, p_s:

$$p_s = \pi_1 + (\pi_2 - \pi_1)c_7,$$

where

$$c_7 = \frac{OA - S_m(\gamma)}{OA - OB}.$$

When the anomaly is zero, the lunar distance is OA, and p_s is equal to π_1. When the anomaly is 180°, the lunar distance is OB, and p_s is equal to π_2. The correction c_7 for anomaly γ is found in column seven of the parallax table opposite the argument $\gamma/2$. The correction for quadrature, found in column eight, is similarly computed. Here the distance OC' is 39;22, whereas OC was 60. The parallax at quadrature, p_q, is

$$p_q = \pi_3 + (\pi_4 - \pi_3)c_8,$$

where

$$c_8 = \frac{OA' - S'_m(\gamma)}{OA' - OB'}.$$

Columns three and five of the parallax table contain the values of π_1 and π_3, while in columns four and six Ptolemy has given the differences $\pi_2 - \pi_1$ and $\pi_4 - \pi_3$. Thus

$$p_s = c_3(z) + c_4(z) \cdot c_7(\gamma/2) \tag{5.1}$$

and

$$p_q = c_5(z) + c_6(z) \cdot c_8(\gamma/2). \tag{5.2}$$

In Figure 5.8, $CFC'D$ is the lunar eccentric circle with center E, the observer is at O, the moon is at F between apogee C and perigee C', and $\angle COF$ is the double elongation. For a given value of the lunar anomaly, the parallax

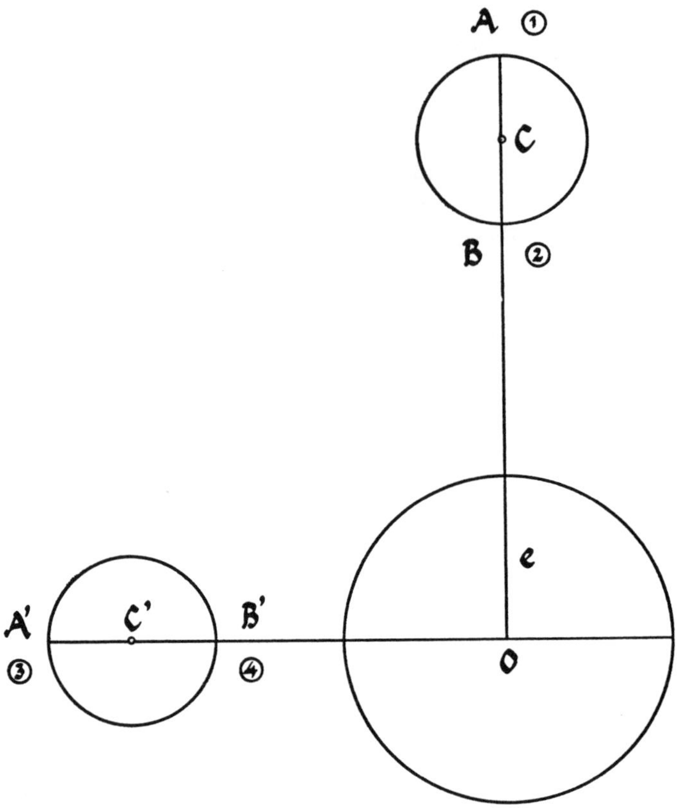

Figure 5.6: Ptolemy's Lunar Parallaxes in Syzygy and Quadrature

varies between a minimum of p_s, and a maximum of p_q. For an intermediate lunar distance OF, Ptolemy uses a factor of interpolation similar to the correction factor of columns seven and eight. If

$$OC = R,$$

$$OE = e,$$

and

$$EF = R' = R - e,$$

the lunar distance OF is

$$S_m(2\eta) = FG + GO = \sqrt{R'^2 - (e \sin 2\eta)^2} + e \cos 2\eta.$$

The final value of the lunar parallax is then

$$p = p_s + (p_q - p_s)c_9(\eta) \tag{5.3}$$

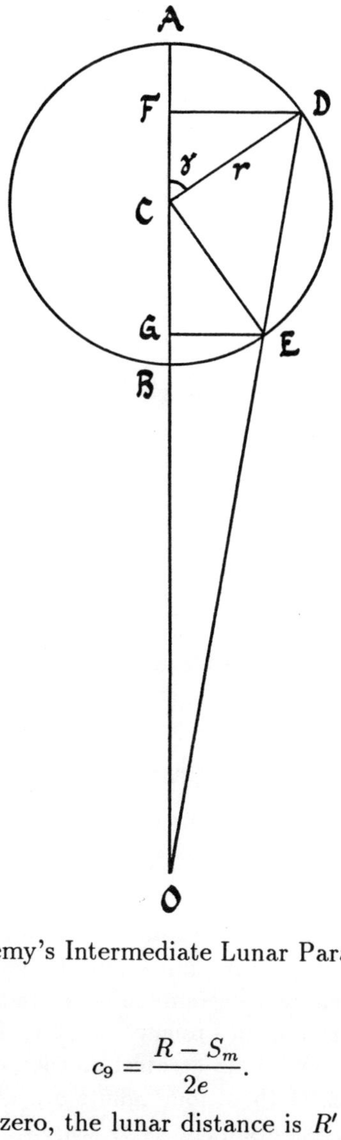

Figure 5.7: Ptolemy's Intermediate Lunar Parallax – Column 7

where

$$c_9 = \frac{R - S_m}{2e}.$$

When the elongation is zero, the lunar distance is $R' + e$, the correction c_9 is zero, and the parallax is equal to p_s. When the elongation is 90°, the lunar distance is $R' - e$, the correction c_9 is one, and the parallax is equal to p_q. c_9 is given in column nine of the parallax table opposite the argument η. Since the greatest distance in quadrature is less than the least distance in syzygy, p_q is always greater than p_s.

4.2 Copernicus' Lunar Parallax

While Copernicus' table of parallax is modeled on Ptolemy's table in the *Almagest*, there are certain differences arising from the nature of their lunar

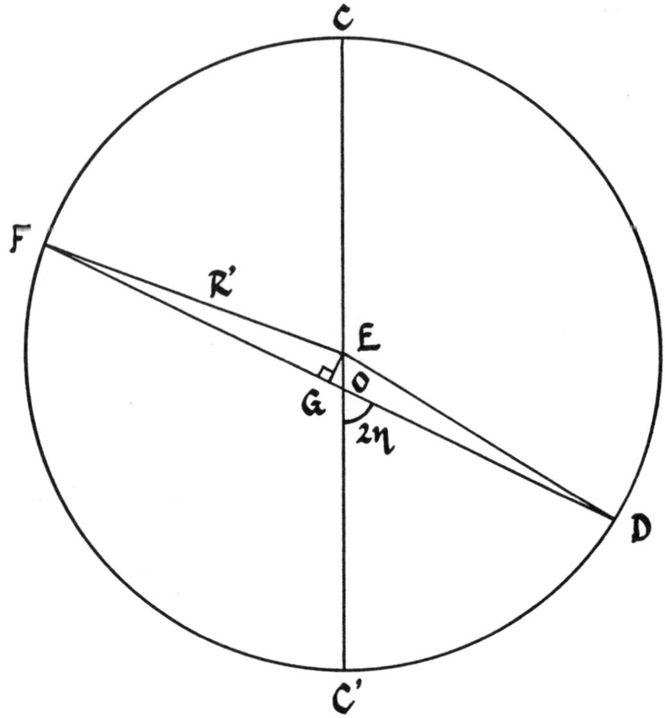

Figure 5.8: Ptolemy's Intermediate Lunar Parallax – Column 9

models. In the Copernican model, there are again two components which affect the lunar distance: the position of the moon on the secondary epicycle, a function of elongation, and the position of the secondary epicycle on the primary epicycle, a function of lunar anomaly. Copernicus considers the two cases, not of syzygy and quadrature, but of the secondary epicycle in apogee or perigee. In Figure 5.9, the four extreme lunar distances have been numbered to correspond to the four cases in Ptolemy, i.e., 1 and 2 are greatest and least distance in syzygy, and 3 and 4 are greatest and least distance in quadrature.

The relative positions of these four points are different for Ptolemy and Copernicus. In addition, the argument is measured from B and A' for Copernicus, but from A and A' for Ptolemy. Thus Copernicus' formulas for the parallax at apogee and perigee, as well as the values in his first four columns, do not correspond to Ptolemy's. In Figure 5.10, D is the center of the earth, AB is the primary lunar epicycle about center C, FGE is the secondary epicycle about the apogee A, and MNO is the position of the secondary epicycle about perigee B. If the moon is at G on the secondary epicycle, with double elongation $\angle EAG$, a correction factor analogous to Ptolemy's would be

$$c_7 = \frac{DG - DF}{DE - DF},$$

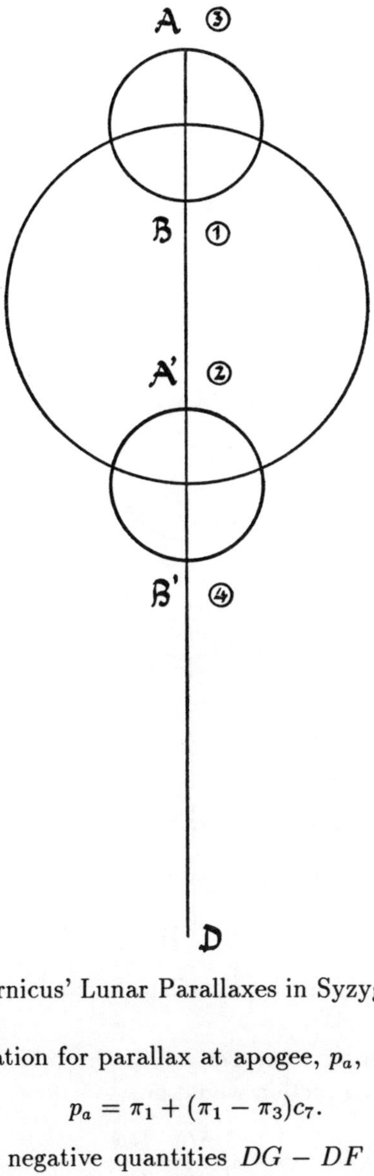

Figure 5.9: Copernicus' Lunar Parallaxes in Syzygy and Quadrature

to be used in the equation for parallax at apogee, p_a,

$$p_a = \pi_1 + (\pi_1 - \pi_3)c_7.$$

In order to avoid the negative quantities $DG - DF$ and $DE - DF$, and in order for zero elongation to correspond to a zero value of c_7, an alternative correction is

$$c_7 = \frac{DG - DE}{DF - DE},$$

to be used in the equation

$$p_a = \pi_1 - (\pi_1 - \pi_3)c_7.$$

Copernicus in theory uses the second alternative. He approximates the distance DG by finding $DC + CG$.

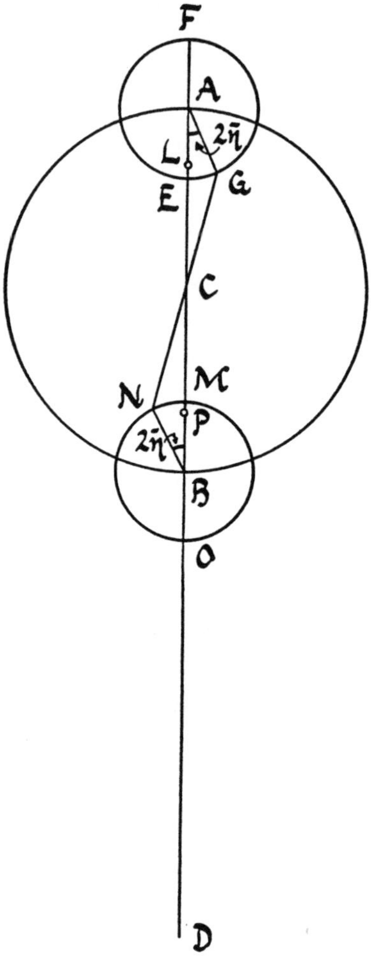

Figure 5.10: Copernicus' Intermediate Lunar Parallax – column 7

For the secondary epicycle at perigee, if the moon is at N with double elongation $\angle MBN$, the correction would be

$$c_8 = \frac{DM - DN}{DM - DO},$$

to be used in the equation for parallax at perigee, p_p,

$$p_p = \pi_2 + (\pi_4 - \pi_2)c_8.$$

The first four columns of lunar parallax in Copernicus' table contain the following:

$$c_3 = \pi_1 - \pi_3,$$

$$c_4 = \pi_1,$$

$$c_5 = \pi_2,$$

and

$$c_6 = \pi_4 - \pi_2.$$

The argument for these columns is twice the zenith distance.

Copernicus gives an example of the computation of c_7 and c_8 for $2\bar{\eta}$ equal to 60°. His results in each case agree to the nearest minute, but he mentions that they actually differ by a few seconds. He gives only one column of correction, i.e., there is no column eight. The greatest difference between columns seven and eight would amount to 42 seconds if computed accurately. Copernicus in fact does not use either of the above formulas for the values in his table. Instead he has inserted the column of proportional parts from his lunar correction table in chapter 11. In Figure 5.11, tangents have been drawn from the earth at D to circles about C, the center of the primary lunar epicycle. The radius CE is the distance from this center to the perigee of the secondary epicycle, CF is the distance to the apogee, and CG is the distance to an arbitrary point on the secondary epicycle. In the lunar correction table, the column of proportional parts is computed as follows:

$$c_4 = \frac{\angle GDC - \angle EDC}{\angle FDE}.$$

The computation of column seven of the parallax table, as indicated by Copernicus' example, would produce

$$\begin{aligned} c_7 &= \frac{(DC + CG) - (DC + CE)}{(DC + CF) - (DC + CE)} \\ &= \frac{CG - CE}{CF - CE} = \frac{\sin \angle GDC - \sin \angle EDC}{\sin \angle FDE}. \end{aligned}$$

Thus the substitution of the column of proportional parts for the correction factor c_7, where the angles themselves have been used in place of the sines, is a reasonable approximation.[20] Copernicus' formulas for the lunar parallax at perigee and apogee are thus

$$p_a = c_4(2z) - c_3(2z)c_7(2\bar{\eta})$$

and

$$p_p = c_5(2z) + c_6(2z)c_7(2\bar{\eta}).$$

The final column of Copernicus' parallax table, column eight, is a correction for the varying distance of the moon as a function of the position of the secondary epicycle on the primary epicycle. In principle, this function is similar to Ptolemy's last column, but again there is a difference arising from the models. Ptolemy's table readily yields a value of the parallax at syzygy (equation (5.1)

[20] Accurately computed, the greatest difference between column seven of the parallax table and column four of the lunar correction table would be $3'$.

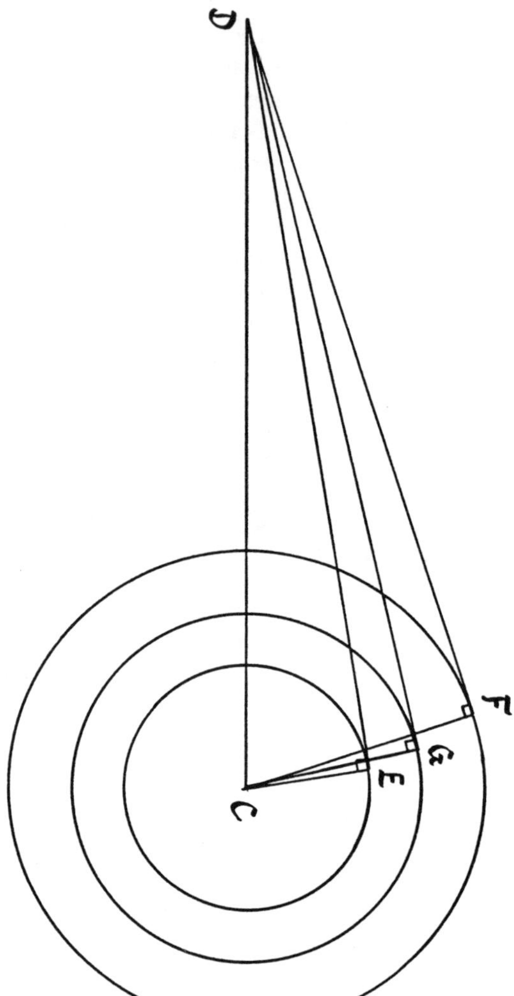

Figure 5.11: Copernicus – Column 7 and the Column of Proportional Parts

above), and his final result is for arbitrary values of elongation and anomaly. If Copernicus had computed the last column of his table using the mean radius of the primary epicycle, his final value of the parallax would also be for arbitrary elongation and anomaly. However, Copernicus uses the radius of the epicycle for syzygy, the distance CE in Figure 5.10. Thus, strictly speaking, the results obtained from Copernicus' table are valid only for syzygy. Copernicus is fully aware of this, as indicated by his remarks at the end of chapter 25, where he explains the use of the parallax table. He says that one needs lunar parallax at syzygy for predicting eclipses, and that the difference between parallax at syzygy and parallax elsewhere is small.[21]

In Figure 5.12, AB is the primary epicycle with radius $5;11^{tr}$, where CD is $60;19^{tr}$. For a point E on the epicycle, with anomaly γ, the value of the interpolation function is

$$c_8 = \frac{AD - AE}{AD - BD}.$$

The argument for column eight is γ. The final formula for lunar parallax is

$$\pi = p_a + (p_p - p_a)c_8(\gamma).$$

Copernicus' examples of the computation of the corrections for apogee and perigee, c_7 and what would have been c_8, contain an inexplicable error. There are two ways in which one may compute these functions. For the epicycle in the apogee of the eccentric (see Figure 5.7), Ptolemy computes the values

$$\frac{OA - OD}{OA - OB},$$

with OC equal to 60 and an epicycle radius of 5;15. In computing column eight he norms OC' to 60 (Figure 5.6), which necessitates multiplying the value of the radius, and thus of BG and EG, by 60/39;22.

An alternate method is to begin with true distances in terrestrial radii, rather than relative distances. This is the procedure followed by Copernicus, and with good reason. In Figure 5.10, his intermediate lunar distances DG and DN are approximated by $DC + CG$ and $DC - CN$. Due to symmetry, when the double elongation angles GAE and NBM are equal, triangles CAG and CBN are congruent. L and P are points such that CG equals CL and CN equals CP. If the numerator of c_7, EL, is calculated with true distances, the same value can then be used for MP in "c_8". The value of EL and MP must then be reduced to account for their different distances from the earth at D. This is where Copernicus' error arises.

Preceding his examples, Copernicus states the following values for the parameters of the model:

$$CE = 5;11^{tr},$$

[21] In fact, the difference between a mean radius of $6;36^{tr}$ and the syzygy radius of $5;11^{tr}$ amounts to only 21″ in column eight. The maximum effect of this on the final value of the parallax is 10″.

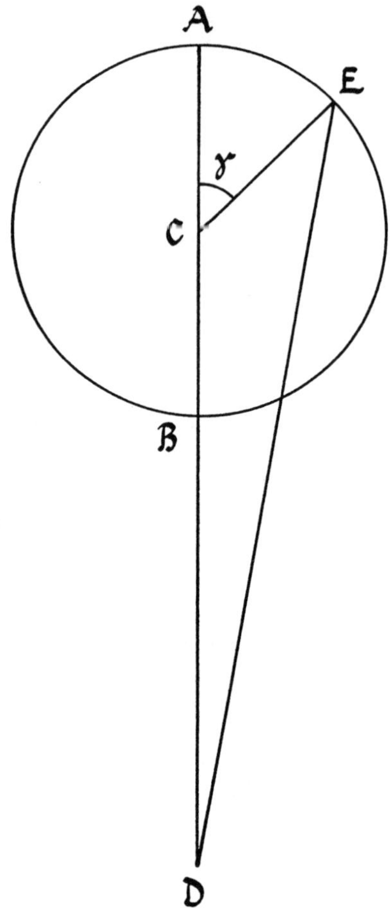

Figure 5.12: Copernicus' Intermediate Lunar Parallax – Column 8

$$DC = 60;18^{tr},$$
$$EF = 2;51^{tr},$$
$$AG = 1;25^{tr},$$
$$AC = 6;36^{tr},$$

and

$$DCE = 65;30^{tr}.$$

The general formula for the length of CG is

$$CG = \sqrt{AG^2 + AC^2 - 2AG \cdot AC \cos 2\bar{\eta}}.$$

In his example, Copernicus uses a double elongation of 60°. The value of CG should then be 6;1,5^{tr}, but Copernicus has 6;7^{tr}. He then finds

$$DCG = DCL = DC + CG = 60;18^{tr} + 6;7^{tr} = 66;25^{tr},$$

and

$$EL = DCL - DCE = 66;25^{tr} - 65;30^{tr} = 0;55^{tr}.$$

Copernicus has written 0;55,30^{tr} for EL.

At this point, Copernicus says that if DCE is 60, EF' is 2;37°, which is correct, since

$$EF' = EF \cdot \frac{60}{DCE} = 2;51^{tr} \cdot \frac{60}{65;30^{tr}} = 2;36,38.$$

However, he also says that EL' is 0;46. This does not follow from any of his values. It does agree, however, with a correctly computed value of EL. Using 60;19^{tr} for DC, instead of 60;18^{tr},[22]

$$DCG = 60;19^{tr} + 6;1,5^{tr} = 66;20,5^{tr},$$

$$EL = 66;20,5^{tr} - 65;30^{tr} = 0;50,5^{tr},$$

and

$$EL' = EL \cdot \frac{60}{DCE} = 0;50,5^{tr} \cdot \frac{60}{65;30^{tr}} = 0;45,52.$$

Finally, the entry in column seven is found by taking

$$\frac{EL'}{EF'} = \frac{0;46}{2;37} = 0;17,35.$$

Copernicus' value is 0;18.

The same error occurs in the second example, for the secondary epicycle at perigee. DBM is taken as 60, rather than 55;8^{tr}, and

$$MP = EL = 0;55,30^{tr}.$$

Copernicus then finds[23]

$$MBO' = 3;7,$$

and

$$MP' = 0;55.$$

Again, the value of MP' does not follow from the values stated by Copernicus, but it does follow from the correct values, namely

$$MP' = MP \cdot \frac{60}{DBM} = 0;50,5^{tr} \cdot \frac{60}{55;8^{tr}} = 0;54,30.$$

---

[22] Copernicus has made very few corrections in the manuscript of this chapter. His original value of DC is clearly 60;18^{tr}, but in computing the last column of the parallax table he states and uses 60;19^{tr}. Just prior to stating the value of EL in his example, he has originally written a 60;19 in place of 65;30. Copernicus' alternation between the use of 60;18 and 60;19 began with the derivation of this quantity in chapter 17.

[23] Accurately,

$$MBO' = 2;51^{tr} \cdot \frac{60}{55;8^{tr}} = 3;6,5.$$

Copernicus states that there is a difference of a few seconds between the corrections for the secondary epicycle at apogee and perigee, as indeed there would be if they were computed correctly. However, if one uses the method given in his examples, the two columns are identical. Column seven is found by taking

$$c_7 = \frac{EL \cdot \frac{60}{65;30}}{EF \cdot \frac{60}{65;30}},$$

and column "eight" by taking

$$c_8 = \frac{MP \cdot \frac{60}{55;8}}{MO \cdot \frac{60}{55;8}}.$$

Since EF and MO are equal, and, due to Copernicus' method of approximating the lunar distance, EL is also equal to MP, the use of the ratios of the relative and true distances to account for the difference between apogee and perigee is meaningless.

Copernicus' computation of an example of his last column is essentially correct. The general formula for the lunar distance DE in Figure 5.12 is

$$DE = \sqrt{CE^2 + DC^2 - 2CE \cdot DC \cos(180^\circ - \gamma)}.$$

For a lunar anomaly of 60°, the correct value of DE is $63;4^{tr}$. Copernicus has originally written $63;3^{tr}$, and uses this value to find

$$DCA - DE = 65;30^{tr} - 63;3^{tr} = 2;27^{tr}.$$

In the manuscript he has corrected the value of DE to $63;4^{tr}$, but he does not bother to change $DBA - DE$ to $2;26^{tr}$.

4.3 Reinhold's Parallax Table

Reinhold's table of parallax in altitude is identical in principle to that of Copernicus. The individual entries differ, of course, because Reinhold uses his recomputed values of solar and lunar distances. One slight difference in computational technique is apparent in column seven. Reinhold solves triangle DAG (see Figure 5.10) for the lunar distance DG, rather than approximating DG as $DC + CG$. Column seven in Reinhold's table is computed for the secondary epicycle in apogee, as described in Copernicus' example. He does not substitute his column of proportional parts from the lunar correction table.

Reinhold's entries are computed with much greater care. In the first five columns of solar and lunar parallax, the discrepancy between Reinhold's values and recomputation by electronic computer is never more than one second. For Copernicus, the discrepancy is as much as five seconds. Reinhold gives both minutes and seconds for columns seven and eight, while Copernicus gives only minutes.[24]

[24] The entry in Reinhold's table in column seven corresponding to 64° is a misprint. The value should read 1<u>6</u>,12, not 1<u>5</u>,12.

4.4 The Diameter Tables

The radii of the sun, moon and shadow vary with the distances of the sun and moon. Copernicus and Reinhold give tables of these radii as a function of solar and lunar anomaly. In chapter 21, Copernicus found a solar radius of 0;15,54° for greatest solar distance (0° of anomaly), and 0;16,57° for least solar distance (180° of anomaly). His column of solar radii begins with 0;15,50° for 6° of anomaly, but does end with 0;16,57° at 180°. The first third of this column is in error by 3 or 4 seconds, but the last third is in error by no more than 1 second. Reinhold gives three columns of solar radii, for minimum, mean and maximum solar eccentricity. His columns begin with zero degrees of anomaly, and all ninety values have been carefully computed, differing in only a few cases by one second from a recomputation by electronic computer.

The third column of Copernicus' table contains the lunar radii in syzygy as a function of anomaly. This column is computed by using the approximation given in chapter 22, that the lunar radius is 2/7 times the lunar parallax. Since it is the horizontal parallax which is used, the zenith distance is 90°, and since the moon is in syzygy, the elongation is 0°. The equation for parallax is then

$$\pi = c_5(180^o) + [c_6(180^o) - c_5(180^o)]c_8(\gamma).$$

Copernicus' parallax table gives

$$c_5(180^o) = 0;52,24^o.$$

The correct value should be 0;52,29°, and this is the value that Copernicus in fact uses in computing the lunar radius column. The formula

$$R_m = \frac{2}{7}[0;52,29^o + 0;9,52^o \cdot c(\gamma)]$$

produces Copernicus' results with an error of no more than one second. Reinhold appears to have actually computed each of the lunar distances in syzygy as a function of anomaly, since the use of this method on the computer reproduces his results exactly.

The fourth column of Copernicus' table contains the shadow radius. In the printed edition this column has been computed by using the equation

$$R_{sh} = \frac{403}{150}R_m.$$

With two exceptions, this method produces Copernicus' results with an error of no more than one second.[25] In the manuscript, Copernicus has computed this column using

$$R_{sh} = \frac{79}{30}R_m,$$

but the values have been calculated with much less care.

---

[25] The two exceptions are at 72°, where the error is 8″, and 114°, where the error is 7″.

Reinhold gives three columns for the shadow radii, labeled minimum, mean and maximum eccentricity. This implies that he has only allowed for one variable, i.e., the shadow radius is computed by assuming a fixed solar distance and varying the lunar distance in syzygy as a function of anomaly. In Figure 5.3, the solar radius is DC, the radius of the earth is KE, and the shadow is MR. The angular shadow radius is

$$\angle RKM = \tan^{-1}\frac{MR}{KM},$$

where KM is the lunar distance, and

$$MR = \frac{MS}{KS} = \frac{KS - KM}{KS}$$

when the terrestrial radius is one. In his commentary on chapter 19 of *De Revolutionibus*, Reinhold computed the values of KS, the shadow axis, for greatest and least solar distance in apogee and perigee. His value of $268;8^{tr}$ for greatest solar distance in apogee, in the case of maximum eccentricity, will produce the results in his third column, when used in the above equations. The first two columns are not as easy to explain. By varying the value of KS, it is possible to reproduce Reinhold's results to within one second. The value of KS which will produce the first column, labeled minimum eccentricity, implies a solar distance of 1179^{tr}.[26] This distance is approximately half-way between the mean solar distance ($1159;40^{tr}$) and the least solar distance in apogee (1197^{tr}). The value of KS which will produce the column labeled mean eccentricity implies a solar distance of 1195^{tr}. This distance is close to the least solar distance in apogee.[27]

The last column in the tables of radii of Copernicus and Reinhold accounts for the variation of the shadow radius. This variation depends on solar distance when the lunar distance is fixed. The column of shadow variation in the Nuremberg edition of *De Revolutionibus* can be produced by assuming a lunar distance of 62^{tr}, and allowing the solar distance to vary from apogee to perigee as a function of solar anomaly. The errors in the first third of this column are similar to those in the column of solar radius, which was also based on solar distances. There is a different set of values in the manuscript, which can be

[26] The solar distance as a function of the shadow axis is

$$DK = KS(DC - 1),$$

where DC is the solar radius of $5;30,20^{tr}$.

[27] While Reinhold uses the tangent to find the angular shadow radius in his commentary, use of the approximation

$$\angle RKM = \sin^{-1}\frac{MR}{KM}$$

produces close agreement with the values in the shadow radius columns. Use of the tangent results in a solar distance of 1182^{tr} for minimum eccentricity.

reproduced by assuming a lunar distance of $65;30^{tr}$. The values in this last column are always to be subtracted from the shadow radius found in the previous column, yielding a corrected shadow radius. I have not been able to determine the value of the lunar distance which will produce Reinhold's column of the variation of the shadow.

APPENDIX A

THE COMPUTATION OF THE LUNAR PARALLAX OBSERVATIONS OF PTOLEMY AND COPERNICUS

This appendix contains details of my own recomputation of the Ptolemaic and Copernican parallax observations. The definitions, methods and formulas are described in chapter one. There are two types of numerical quantities in what follows: quantities for which Ptolemy, Copernicus and Reinhold give their own values, and those for which they do not. If a numerical value is not specifically identified with Ptolemy, Copernicus or Reinhold, then one may safely assume that it is my own recomputed value. In general it is also true that if I do not point out a discrepancy between my recomputed values and those in the *Almagest*, *De Revolutionibus* or Reinhold's manuscripts, one may assume that they are in agreement. One can easily ascertain which quantities are given by the three astronomers and what their values are by referring to Tables 1.3 to 1.10 in chapter one. Bracketed numbers in the tables are not given by Ptolemy, Copernicus and Reinhold but have been reconstructed from the existing values.

My recomputation of Ptolemy's values has been done using the tables in the *Almagest*. For Copernicus, I have used the Nuremberg edition of *De Revolutionibus*, and for Reinhold, the *Prutenic Tables* (1551). In computing the mean motions, I have truncated the values in the mean motion tables of Ptolemy at the third sexagesimal place. Ptolemy and Copernicus give only the final results of their computations of the mean motions, but Reinhold's manuscript contains the individual values for years, days and hours. Reinhold's practice is to raise by one the third sexagesimal place of the value in the *Almagest* mean motion tables when severing a fourth sexagesimal place of 30 or more.

PTOLEMY'S PARALLAX OBSERVATION
(Table 1.3)

Ptolemy's lunar parallax observation was made in Alexandria on Athyr 13, Hadrian 20 (October 1, 135), $5;50^h$ after noon, when the zenith distance of the moon was $50;55^o$. The time since epoch was $882^a\ 72^d\ 5;50^h$.

Solar Longitude and the Equation of Time

The solar anomaly at the time of the observation is

	$\bar{k}$
epoch	265;15
810^a	163; 4,12,15
72^a	342;29,42,25
60^d	59; 8,17,13
12^d	11;49,39,26
5^h	0;12,19,13
$0;50^h$	0; 2, 3,12
	122; 1,13,44.

The equation of center is then

$$c_s(122;1,14^o) = 2;3,18^o,$$

the mean longitude of the sun is

$$\bar{\lambda}_s = 122;1,14^o + 65;30^o = 187;31,14^o,$$

and the true solar longitude is

$$\lambda_s = 187;31,14^o - 2;3,18^o = 185;27,56^o.$$

Using Ptolemy's table of right ascension, the equation of time is

$$\begin{aligned} E^o &= \alpha(185;28^o) - \alpha_o - (\bar{\lambda}_s(t) - \bar{\lambda}_s(t_o)) \\ &= 185;1^o - 335;7^o - (187;31^o - 330;45^o) \\ &= -6;52^o. \end{aligned}$$

Converting from equatorial degrees to hours,

$$E^h = 0;4^{h/o} \cdot (-6;52^o) = -0;27,28^h.$$

The corrected time since epoch should thus be 882^a 72^d $5;23^h$. Ptolemy has $5;20^h$.

Lunar Longitude, Latitude and Declination

The mean lunar motions for Ptolemy's corrected time ($5;20^h$) are as follows:

	$\bar{\lambda}_m$	$\bar{\gamma}$	$\bar{w}$	$\bar{\eta}$
epoch	41;22	268;49	354;15	70;37
810^a	37;24, 6,52	222;10,57,16	217;37,21,59	234;19,54,36
72^a	315;19,28,36	267;44,58,25	267;20,39,17	332;49,46,11
60^d	70;34,58,33	63;53,56,17	73;45,39,48	11;26,41,20
12^d	158; 6,59,42	156;46,47,15	158;45, 7,57	146;17,20,16
5^h	2;44,42,17	2;43,18,44	2;45,22, 0	2;32,23, 3
$0;20^h$	0;10,58,49	0;10,53,14	0;11, 1,28	0;10, 9,32
	265;43,14,49	262;19,51,11	354;40,12,29	78;13,14,58.

Ptolemy's value of $\bar{\lambda}_m$ is 265;44°, rather than 265;43°. He apparently has found

$$\begin{aligned}\bar{\lambda}_m(882^a\ 72^d\ 5;20^h) &= \bar{\eta}(882^a\ 72^d\ 5;20^h) + \bar{\lambda}_s(882^a\ 72^d\ 5;\underline{50}^h)\\ &= 78;13^o + 187;31^o\\ &= 265;44^o.\end{aligned}$$

The lunar corrections and true anomaly are

$$\begin{aligned}c_3(156;26^o) &= 7;40^o,\\ \gamma &= 262;20^o + 7;40^o = 270;0^o,\\ c_4(270;0^o) &= 4;59^o,\\ c_5(270;0^o) &= 2;35^o,\end{aligned}$$

and

$$c_6(156;26^o) = 0;56,42^o.$$

The lunar equation of center is

$$c_m = 4;59^o + 2;26,28^o = 7;25,28^o.$$

Using Ptolemy's value of 7;26°, the true lunar longitude is

$$\lambda_m = 265;44^o + 7;26^o = 273;10^o,$$

and the true argument of latitude is

$$w = 354;40^o + 7;26^o = 2;6^o.$$

The lunar latitude, taken from the last column of the correction table, is then

$$\beta(2;6^o) = 4;59^o$$

north of the ecliptic. The declination of the longitude of the moon, from Ptolemy's table of declination, is

$$\delta^*_m\ (273;10^o) = 23;48,59,50^o.$$

Ptolemy has 23;49°.

REINHOLD'S RECOMPUTATION OF PTOLEMY'S OBSERVATION (Table 1.3)

Solar Longitude and the Equation of Time

Reinhold's manuscript contains the following values for the mean motion in solar anomaly:

	$\bar{k}$
epoch	265;15, 0, 0
810^a	163; 4,12,15
72^a	342;29,42,25
60^d	59; 8,17,13
12^d	11;49,39,27
5^h	0;12,19,14
$0;50^h$	0; 2, 3,12
	122; 1,13,46

Then

$$\bar{\lambda}_s = 122;1,13,46^o + 65;30^o = 187;31,13,46^o,$$

which Reinhold rounds to 187;31,14°,

$$c_s(122;1,14^o) = 2;3,18^o,$$

for which he has 2;3°, and

$$\lambda_s = 187;31,14^o - 2;3^o = 185;28,14^o,$$

which he writes as 185;28°. Since Reinhold's values of $\bar{\lambda}_s$ and λ_s agree with Ptolemy's to one sexagesimal place, Reinhold's equation of time should also be $-0;27^h$. He does not calculate it himself, but uses Ptolemy's value of −30 minutes.

Lunar Longitude, Latitude and Declination

Reinhold finds the mean lunar longitude, rather than the elongation, in addition to the mean lunar anomaly and argument of latitude:

	$\bar{\lambda}_m$	$\bar{\gamma}$	$\bar{\varpi}$
epoch	41;22, 0, 0	268;49, 0, 0	354;15, 0, 0
810^a	37;24, 6,52	222;10,57,17	217;37,22, 0
72^a	315;19,28,37	267;44,58,25	267;20,39,17
60^d	70;34,58,34	63;53,56,18	73;45,39,48
12^d	158; 6,59,43	156;46,47,16	158;45, 7,58
5^h	2;44,42,17	2;43,18,44	2;45,22, 1
$0;20^h$	0;10,58,49	0;10,53,15	0;11, 1,28
	265;43,14,52	262;19,51,15	354;40,12,32.

Since the value of $\bar{\varpi}$ in the *Almagest* for 60^d is 73;45,39,48,56,37,0°, one would expect Reinhold to round this to 73;45,39,4<u>9</u>°. Reinhold appears to have found the mean lunar elongation correctly, i.e., he has not combined values calculated for different times. Using the corrected time for both solar and lunar longitude,

$$\begin{aligned}\bar{\eta}(882^a\ 72^d\ 5;20^h) &= \bar{\lambda}_m(882^a\ 72^d\ 5;20^h) - \bar{\lambda}_s(882^a\ 72^d\ 5;\underline{20}^h)\\ &= 265;43,14,52^o - 187;29,59,51^o\\ &= 78;13,15,1^o,\end{aligned}$$

whereas

$$\begin{aligned}\bar{\eta}(882^a\ 72^d\ 5;20^h) &= \bar{\lambda}_m(882^a\ 72^d\ 5;20^h) - \bar{\lambda}_s(882^a\ 72^d\ 5;\underline{50}^h)\\ &= 265;43,14,52^o - 187;31,13,46^o\\ &= 78;12,1^o.\end{aligned}$$

Reinhold's value is 78;13,0°.

The lunar corrections and true anomaly are

$$c_3(156;26,0^o) = 7;40^o,$$
$$\gamma = 262;19,51^o + 7;40^o = 269;59,51^o,$$
$$c_4(269;59,51^o) = 4;59^o,$$
$$c_5(269;59,51^o) = 2;35^o,$$

and

$$c_6(156;26,0^o) = 0;56,42^o.$$

The lunar equation of center is

$$c_m = 4;59^o + 2;26,29^o = 7;25,29^o.$$

The true lunar longitude is

$$\lambda_m = 265;43,15^o + 7;25,29^o = 273;8,44^o,$$

and the true argument of latitude is

$$w = 354;40,13^o + 7;25,29^o = 2;5,42^o.$$

Reinhold, however, has written 2;5,22° for w. This does not affect the value of the lunar latitude, which is 4;59°.

Reinhold does not find the lunar declination, either δ_m or δ_m^*. Rather, he finds DE in figure 1.8, the great circle arc through the moon (G) between the ecliptic and equator, perpendicular to the ecliptic. In spherical triangle EDB, with $\angle DBE$, the eccentricity, equal to 23;51,20°, and using Ptolemy's lunar longitude, 273;10°, so that AD is 3;10°,

$$\begin{aligned}\tan DE &= \sin BD / \tan(90^o - \epsilon)\\ &= \sin 86;50^o / \tan 66;8,40^o\\ &= 0.4415361,\end{aligned}$$

and

$$DE = 23;49,23^o.$$

This is Reinhold's result. Reinhold's value of λ_m, 273;8,44°, or the rounded value 273;9°, will not produce 23;49,23°.

MEAN MOTIONS AND EPOCH VALUES FOR COPERNICUS AND REINHOLD

Copernican Mean Lunar Motions

In *De Revolutionibus* iv.4, Copernicus gives the following values for the mean lunar motions in an Egyptian year:

$$\bar{\eta}_a = 129;37,22,32,40^o,$$

$$\bar{\gamma}_a = 88;43,9,5,9^o,$$

and

$$\bar{\varpi}_a = 148;42,44,31,6^o.$$

This last value, however, has been changed in the manuscript to read

$$\bar{\varpi}_b = 148;42,45,17,21^o.$$

Directly following these values are the tables themselves. The mean motion tables were initially computed for the first set of values, $\bar{\eta}_a$, $\bar{\gamma}_a$ and $\bar{\varpi}_a$. These parameters were subsequently altered, as can be seen by the corrections made in the tables. The elongation and anomaly have first been changed to

$$\bar{\eta}_b = 129;37,22,36,15^o$$

and

$$\bar{\gamma}_b = 88;43,9,13,45^o,$$

and then to

$$\bar{\eta}_c = 129;37,22,36,25^o$$

and

$$\bar{\gamma}_c = 88;43,9,7,15^o.$$

The values in the table for the mean argument of latitude have been corrected only once, from $\bar{\varpi}_a$ to $\bar{\varpi}_b$.

True Solar Longitude at Epoch – Copernicus

In attempting to recompute Copernicus' value for the equation of time, there are two elements of uncertainty. One concerns the variation in the obliquity of the ecliptic. For purposes of computing the right ascensions used in the equation of time, very little error is introduced if one assumes Copernicus' value of the maximum obliquity (23;52°) at epoch, and his minimum obliquity (23;28°) in the early sixteenth century.

Secondly, Copernicus does not state his value for the true solar longitude at epoch. This value, and its right ascension, can be calculated using Copernicus' epoch values for the precession, the anomaly of the equinoxes, the simple sun and the anomaly of the sun. At epoch

$$\begin{aligned} \bar{P}(t_o) &= 5;32^o, \\ \theta(t_o) &= 6;45^o, \\ 2\theta(t_o) &= 13;30^o, \\ \lambda_s^*(t_o) &= 272;31^o, \\ \bar{l}_s^*(t_o) &= 278;\ 2^o, \end{aligned}$$

and

$$\bar{k}(t_o) = 211;19^o.$$

From the correction table for the equinox and obliquity

$$c_p(13;30^o) = 0;16^o.$$

Thus

$$P(t_o) = 5;32^o + 0;16^o = 5;48^o,$$

and

$$\bar{\lambda}_s(t_o) = 272;31^o + 5;48^o = 278;19^o.$$

The epoch value of the composite sun should be

$$\bar{l}_s(t_o) = 272;31^o + 5;32^o = 278;3^o,$$

but Copernicus has 278;2°, i.e., his epoch values for $\bar{P}$, λ_s^* and $\bar{l}_s$ do not agree. The solar corrections and true anomaly are

$$c_3(6;45^o) = 0;46,15^o,$$

$$k = 211;19^o + 0;46,15^o = 212;5,15^o,$$

$$c_4(212;5,15^o) = 1;1,29^o,$$

$$c_5(212;5,15^o) = 0;17,42^o,$$

and

$$c_6(6;45^o) = 0;60^o = 1;0^o.$$

The equation of center is

$$c_s = 1;1,29^o + 0;17,42^o = 1;19,11^o,$$

and thus the true solar longitude at epoch is

$$\lambda_s(t_o) = 278;19^o + 1;19,11^o = 279;38,11^o.$$

Using Copernicus' table of right ascension, and assuming maximum obliquity,

$$\alpha_o(279;38,11^o) = 280;31^o,$$

the right ascension of the solar longitude at epoch.

Epoch values – Reinhold

The epoch values given in the *Prutenic Tables* are for the latitude of Königsberg. In the commentary on *De Revolutionibus*, however, the epoch positions are computed for Frauenburg. Reinhold uses the latter in recomputing the observations of Copernicus. The values of these quantities, and their location in the manuscript, are as follows:

$\lambda_s^*(t_o)$	=	272;30, 3,52,7,25°	(Ms.R.16r)
$\bar{l}_s(t_o)$	=	278; 2,27,59,8,25°	(Ms.R.16v)
$\bar{k}(t_o)$	=	211;39,14,40,1,8°	(Ms.R.23v)
$\theta(t_o)$	=	6;40,27°	(Ms.R.27v)
$\bar{\eta}(t_o)$	=	210; 0,55°	(Ms.R.40v)
$\bar{\gamma}(t_o)$	=	207;16,11°	(Ms.R.40v)
$\bar{\varpi}(t_o)$	=	129;44,35,40°	(Ms.R.40v).

The mean motion and epoch values of the precession and anomaly of the equinoxes are not given in the manuscript, but the epoch position of the precession can be recovered from those of the simple and composite sun:

$$\bar{P}(t_o) = \bar{l}_s(t_o) - \lambda_s^*(t_o) = 5;32,24,7,1,0^o.$$

Reinhold rounds all values to two sexagesimal places.

True Solar Longitude at Epoch – Reinhold

The elements of uncertainty in Copernicus' equation of time are not present in Reinhold's computation. At the end of the commentary on book three of *De Revolutionibus* (Ms.R.27v), Reinhold derives the necessary quantities. His values are

$$\bar{l}_s(t_o) = 278;\ \ 2,28^o,$$

$$\lambda_s(t^o) = 279;\ \ 5,24^o,$$

and the right ascension of $\lambda_s(t_o)$, assuming maximum obliquity of the ecliptic, as Reinhold explicitly states, is

$$\alpha_o = 279;55,33^o.$$

THE FIRST PARALLAX OBSERVATION OF COPERNICUS (Tables 1.4, 1.5)

The date of this observation is the fifth day before the Kalends of October (September 27), 1522. Copernicus wavered in his choice of a time of day. The recomputation of his solar longitude has been done here under the assumption that the apparent time of the observation was 5;40^h after noon. The time since epoch is then 1522^a 284^d 17;40^h. The apparent lunar zenith distance was 82;50°.

Solar Longitude and the Equation of Time

The mean precession and anomaly of the equinoxes at the time of the observation are

	$\bar{p}$	θ
epoch	5;32	6;45
1500^a	20;55, 2	2,37;15, 3
22^a	0;18,24,25	2;18,22,51
240^d	0; 0,33	0; 4, 8
44^d	0; 0, 6, 3	0; 0,45,29
17;40^h	0; 0, 0, 6	0; 0, 0,46
	26;46, 5,34	2,46;23,20, 6

The correction of the precession is

$$c_p(332;46,40^o) = 0;32,13^o,$$

and thus the true precession is

$$P = 26;46,5^o + 0;32,13^o = 27;18,18^o,$$

Copernicus may have used either the simple sun or the composite sun to compute the solar longitude. Since Reinhold uses the simple sun, the same is done here for comparison.

	λ_s^*	$\bar{k}$
epoch	4,32;31	3,31;19
1500^a	5;40;27,56	5,30;19,30
22^a	5,54;26, 0,35	5,54;17, 5,10
240^d	3,56;32,45	3,56;32,29
44^d	43;22, 0,20	43;21,57,24
17^h	0;41,53,18, 3	0;41,53,15,12
0;40^h	0; 1,38,33,39	0; 1;38,33,32
	2,48; 3,13,46,42	1,36;33,33,22,44

The mean solar longitude is

$$\bar{\lambda}_s = 168;3,14^o + 27;18,18^o = 195;21,32^o,$$

and the composite solar longitude should be

$$\bar{l}_s = 168;3,14^o + 26;46,5^o = 194;49,19^o.$$

Since Copernicus' epoch values for $\bar{P}$, λ_s^* and $\bar{l}_s$ do not agree, i.e.,

$$\bar{l}(t_o) \neq \lambda_s^*(t_o) + \bar{P}(t_o),$$

if the composite solar longitude is computed from the mean motion tables, and the epoch value of $\bar{l}_s$, the result is 194;48,19°. This is the value used below.

The solar corrections and true anomaly are

$$c_3(166;23,20^o) = 2;1,34^o,$$
$$k = 96;33,33^o + 2;1,34^o = 98;35,7^o,$$
$$c_4(98;35,7^o) = 1;50^o,$$
$$c_5(98;35,7^o) = 0;32,8^o,$$

and

$$c_6(166;23,20^o) = 0;1^o.$$

The solar equation of center is

$$c_s = 1;50^o + 0;0,32^o = 1;50,32^o,$$

and the true solar longitude is

$$\lambda_s = 195;21,32^o - 1;50,32^o = 193;31,0^o.$$

Copernicus finds first 193;34° and then 193;29° for λ_s (see Table 1.4).

Using my own computed value of λ_s, and assuming minimum obliquity of the ecliptic at the time of the observation, the equation of time is

$$\begin{aligned} E^o &= \alpha(193;31^o) - \alpha_o - (\bar{l}_s(t) - \bar{l}_s(t_o)) \\ &= 192;25,25^o - 280;31^o - (194;48,19^o - 278;2^o) \\ &= -4;51,54^o, \end{aligned}$$

and

$$E^h = 0;4^{o/h} \cdot (-4;51,54^o) = -0;19,28^h.$$

Copernicus found $-0;16^h$.

Lunar Longitude and Latitude

The mean lunar motions at the time of the observation, calculated for the different values found in the manuscript, are listed in Table 1.5. They were obtained as follows (only those values which are affected by the changes are listed in the second and third columns):

	$\bar{\eta}_a$	$\bar{\eta}_b$	$\bar{\eta}_c$
epoch	3,29;58		
1500^a	34;23,26	34;25, 6	34;25,10
22^a	5,31;42,15,58	5,31;42,17,18	5,31;42,17,21
240^d	45;46,46		
44^d	2,56;23,34,26		
17^h	8;38, 6,24		
$0;24^h$	0;12,11,27		
	1,27; 4,30,15	1,27; 6, 1,35	1,27; 6, 5,38

	$\bar{\gamma}_a$	$\bar{\gamma}_b$	$\bar{\gamma}_c$
epoch	3,27; 7		
1500^a	3,58;47, 8	3,58;50,44	3,58;48, 1
22^a	2,31;49,19,53	2,31;49,23, 3	2,31;49,20,39
240^d	4,15;35,46		
44^d	3,34;51,33,26		
17^h	9;15,15,43		
$0;24^h$	0;13, 3,54		
	5,57;39, 6,56	5,57;42,46, 6	5,57;40, 0,42

	$\bar{w}_a$	$\bar{w}_b$
epoch	2, 9;45	
1500^a	3,48;32,57	3,48;52,13
22^a	31;40,19,24	31;40,36,21
240^d	4,55; 2,37	
44^d	3,42; 5,28,52	
17^h	9;22,14,50	
$0;24^h$	0;13,13,46	
	3,16;41,50,52	3,17; 1,23,49

The lunar position below has been computed from Copernicus' values

$$\bar{\eta} = 87;\ 6^o,$$

$$\bar{\gamma} = 357;39^o,$$

and

$$\bar{w} = 197;1^o.$$

The lunar correction c_3 and true anomaly are

$$c_3(174;12) = 1;1,52^o$$

and

$$\gamma = 357;339^o + 1;1,52^o = 358;40,52^o.$$

Copernicus has 358;29°, changed to 358;40°, for γ. Then

$$c_4(358;40^o) = 0;6,13^o,$$

$$c_5(358;40^o) = 0;3,7^o,$$

and

$$c_6(174;12^o) = 0;60^o = 1;0^o.$$

The lunar equation of center is

$$c_m = 0;6,13^o + 0;3,7^o = 0;9,20^o,$$

Copernicus' value, however, is 0;7°. The true elongation of the moon from the mean solar longitude is

$$\eta = 87;6^o + 0;7^o = 87;13^o.$$

Copernicus has two values for the true lunar longitude: first 282;37°, and then 282;32°. The difference, as in the case of the two solar longitudes, is 0;5°. The two values can be explained in two ways. The solar equation of center found above was 1;50,32°. Using the value 1;50° and Copernicus' two values of λ_s,

$$\lambda_m = \lambda_s + c_s + \eta,$$

$$\lambda_m = 193;34^o + 1;50^o + 87;13^o = 282;37^o,$$

or

$$\lambda_m = 193;29^o + 1;50^o + 87;13^o = 282;32^o.$$

If, however, Copernicus used 1;51 for c_s and adjusted the solar longitude from apparent to corrected time, then

$$\lambda_m = 193;33^o + 1;51^o + 87;13^o = 282;37^o$$

or

$$\lambda_m = 193;28^o + 1;51^o + 87;13^o = 282;32^o,$$

since the solar motion in 0;16^h is approximately one minute.

Copernicus' two values of w follow from his second and third values of $\bar{w}$ in the manuscript:

$$w = 194;49^o + 0;7^o = 194;56^o$$

and

$$w = 197;1^o + 0;7^o = 197;8^o.$$

Copernicus has two values for the lunar latitude, 4;50° and 4;47°. Using his two values of w and the correction table,

$$\beta(194;56^o) = 4;49,56^o$$

and

$$\beta(197;8^o) = 4;46,27^o.$$

In the second case, if the latitude is computed exactly, i.e.,

$$\sin\beta = \sin 5^o \cdot \sin(270^o - 197;8^o),$$

the result is 4;47°.

The Declination of the Moon.

In this observation it is not clear whether Copernicus found the declination of the true lunar position (δ_m) or the sum of the declination of the point on the ecliptic which has the same longitude as the moon (δ_m^*) and the latitude. Copernicus gives only one value for the lunar declination, 27;41°. This is equal to the declination of the true lunar position (δ_m), if one assumes Copernicus' uncorrected values for the longitude and latitude. However, since he appears to use $\delta_m^* + \beta$ in the second observation, it may be that he did the same here. The results obtained by using the various alternatives are described in what follows.

In Figure 1.10a, the equator AB, the ecliptic AC, and the lunar orbit AD have been projected onto a plane. The moon is at D. CD, the lunar latitude, is perpendicular to the ecliptic, and DE, the declination (δ_m), is perpendicular to the equator. Assuming Copernicus' final results for the lunar position ($\lambda_m = 282;32^o$, $\beta = 4;47^o$),

$$CD = 4;47^o$$

and

$$AC = 360^o - 282;32^o = 77;28^o.$$

In right triangle ABC, if ∠CAB is the minimum obliquity (23;28°),

$$\tan BC = \sin AC/\tan(90^o - \angle CAB) = 0.42378,$$

and

$$BC = 22;58^o.$$

Then

$$BD = 22;58^o + 4;47^o = 27;45^o.$$

In the same triangle,

$$\cos \angle CBA = \cos AC \cdot \sin \angle CAB = 0.08642,$$

and

$$\angle CBA = 85;2^o.$$

In right triangle BDE,

$$\sin DE = \sin \angle CBA \cdot \sin BD = 0.46386,$$

and the declination is

$$DE = \delta_m = 27;38^o.$$

Using Copernicus' first set of values for lunar longitude and latitude, 282;37° and 4;50°,

$$CD = 4;50^o$$

and

$$AC = 360^o - 282;37^o = 77;23^o.$$

In right triangle ABC,

$$\tan BC = \sin AC / \tan(90^o - \angle CAB) = 0.42364,$$

and

$$BC = 22;58^o,$$

as before. Then

$$BD = 22;58^o + 4;50^o = 27;48^o.$$

In the same triangle

$$\cos \angle CBA = \cos AC \cdot \sin \angle CAB = 0.08698,$$

and

$$\angle CBA = 85;1^o.$$

In right triangle BDE,

$$\sin DE = \sin \angle CBA \cdot \sin BD = 0.46461,$$

and the declination is

$$DE = \delta_m = 27;41^o,$$

which is Copernicus' value.

From Copernicus' table of declination, assuming minimum obliquity, and using the first set of lunar values,

$$\delta^*_m(282;37^o) + \beta = 22;51,55^o + 4;50^o = 27;41,55^o.$$

With the second set,

$$\delta^*_m(282;32^o) + \beta = 22;52,20^o + 4;47^o = 27;39,20^o.$$

The Parallax according to Ptolemy

Copernicus states that the parallax for this observation, computed from Ptolemy's parallax table, would have been 1;17°. The symbols c_3 through c_9, whether they refer to the tables of Ptolemy, Copernicus or Reinhold, represent columns 3 through 9 of the parallax tables in what follows. Using the values

$$z = 82^o,$$

$$\gamma = 358;40^o,$$

and

$$\bar{\eta} = 87;6^o,$$

one finds from Ptolemy's table

$$\begin{aligned} P_1 &= c_3(z) + c_7(\gamma/2) \cdot c_4(z) \\ &= 0;53,9^o + 0;0,5^o \cdot 0;10,4^o = 0;53,10^o, \\ P_2 &= c_5(z) + c_8(\gamma/2) \cdot c_6(z) \\ &= 1;17,40^o + 0;0,4^o \cdot 0;24,20^o = 1;17,42^o, \end{aligned}$$

and the parallax is

$$\begin{aligned} \pi &= P_1 + c_9(\bar{\eta}) \cdot (P_2 - P_1) \\ &= 0;53,10^o + 0;59,45^o \cdot 0;24,32^o = 1;17,36^o. \end{aligned}$$

Copernicus' tables give the following:

$$\begin{aligned} P_1 &= c_3(2z) - c_7(2\bar{\eta}) \cdot c_4(2z) \\ &= 0;52,2^o - 0;60^o \cdot 0;2,5^o = 0;49,57^o, \\ P_2 &= c_5(2z) + c_7(2\bar{\eta}) \cdot c_6(2z) \\ &= 0;61,54^o + 0;60^o \cdot 0;3,23^o = 0;65,17^o, \end{aligned}$$

and

$$\begin{aligned} \pi &= P_1 + c_8(\gamma) \cdot (P_2 - P_1) \\ &= 0;49,57^o + 0^o \cdot 0;15,20^o = 0;49,57^o. \end{aligned}$$

This compares well with Copernicus' value of 0;50°.

REINHOLD'S RECOMPUTATION OF COPERNICUS' FIRST OBSERVATION (Tables 1.6, 1.7)

The time since epoch in Julian years is 1521^j 269^d $17;40^h$.

Solar Longitude and the Equation of Time

Reinhold has found, for the precession and anomaly of the equinoxes:

	$\bar{P}$	θ
epoch	5;32,24	6;40,27
1500^j	20;55,54	2,37;21,31
21^j	0;17,35	2;12,11
269^d	0; 0,37	0; 4,38
	26;46,30	2,46;18,47.

He has omitted the hourly motion, which would have been 0;0,0,6° for $\bar{P}$, and 0;0,0,46° for θ. From Reinhold's correction table

$$c_p(332;37,34^o) = 0;32,49^o.$$

Then

$$P = 26;46,30^o + 0;32,49^o = 27;19,19^o.$$

Reinhold finds the simple sun and solar anomaly:

	λ_s^*	$\bar{k}$
epoch	4,32;30, 4	3,31;39,15
1500^j	5,50; 5,33	5,39;26,17
21^j	5,59;36,54	5,59;27,56
269^d	4,25; 7,53	4,25; 7,24
17^h	0;41,53	0;41,53
$0;40^h$	0; 1,38	0; 1,38
	2,48; 3,45	1,36;24,23

He proceeds directly to the computation of the solar equation:

$$c_3 = (166;18,47^o) = 1;59,35^o,$$

for which he has 2;0°,

$$k = 96;24,23^o + 2;0^o = 98;24,23^o,$$

$$c_4 = (98;24,23^o) = 1;49,57^o,$$

$$c_5(98;24,23^o) = 0;32,39^o,$$

$$c_6(166;18,47^o) = 0;0,58,30^o,$$

for which he has 0;0,58°, and finally

$$c_s = 1;49,57^o + 0;0,31,34^o = 1;50,28,34^o,$$

which he rounds to 1;50,29°. Reinhold finds the true solar longitude by first taking

$$\lambda_s^* - c_s = 168;3,45^o - 1;50,29^o = 166;13,16^o,$$

and then adding the true precession

$$\lambda_s = 166;13,16^o + 27;19,19^o = 193;32,35^o.$$

The composite sun is

$$\bar{l}_s = 168;3,45^o + 26;46,30^o = 194;50,15^o.$$

From Reinhold's table of right ascension

$$\alpha(193;32,35^o) = 192;27,35^o.$$

He uses 192;28° in the equation of time:

$$\begin{aligned} E^o &= 192;28^o - 279;55,33^o - (194;50,15^o - 278;2,28^o) \\ &= -4;15,20^o \end{aligned}$$

and

$$E^h = 0;4^{h/o} \cdot (-4;15,20^o) = -0;17,1,20^h.$$

Reinhold uses –0;17^h. Thus the corrected time since epoch is 1521^j 269^d 17;23^h.

Lunar Longitude and Latitude

Reinhold finds the following mean lunar motions:

	$\bar{\eta}$	$\bar{\gamma}$	$\bar{w}$
epoch	3,30; 0,55	3,27;16,11	2, 9;44,36
1500^j	4,45;53,11	1,37;44,32	2,29;54,37
21^j	4,23; 2, 5	2, 8;25,20	5, 9; 6,41
269^d	39;18,40	4,34;28,50	5,18;41,42
17^h	8;38, 6	9;15,16	9;22,15
0;23^h	0;11,49	0;12,31	0;12,41
	1,27; 4,38	5,57;22,40	3,17; 2,32

Both from accurate computation and from interpolation in Reinhold's tables, the value of the lunar correction, c_3(174;9,16°), should be 1;2,9°. Reinhold writes and uses 1;2,0°. Using this value,

$$\gamma = 357;22,40^o + 1;2,0^o = 358;24,40^o,$$

$$c_4(358;24,40^o) = 0;7,33^o,$$

$$c_5(358;24,40^o) = 0;3,40^o,$$

and

$$c_6(174;9,16^o) = 0;59,52^o.$$

Reinhold does not state his values of c_4 and c_5, and the above values do not produce his result for the lunar equation:

$$c_m = 0;7,33^o + 0;3,40^o = 0;11,13^o.$$

Reinhold has 0;7,35°. The elongation of the true lunar position from the mean solar position is

$$\eta = 87;4,38^o + 0;7,35^o = 87;12,13^o.$$

Reinhold finds the mean solar longitude for the apparent time

$$\bar{\lambda}_s = 168;3,45^o + 27;19,19^o = 195;23,4^o,$$

and combines it with the elongation to obtain the true lunar longitude

$$\lambda_m = 195;23,4^\circ + 87;12,13^\circ = 282;35,17^\circ.$$

The true argument of latitude is

$$w = 197;2,32^\circ + 0;7,35^\circ = 197;10,7^\circ.$$

From the table of lunar latitude

$$\beta(197;10,7^\circ) = 4;46,36^\circ.$$

This is Reinhold's value.

The Declination of the Moon

Reinhold supplies a number of details for his computation of the lunar declination δ_m. The following is a transcription of what appears in his manuscript (Ms.R.50v).

Supputatio declinationis
ex nostris tabulis

Arcus Mer[idianus]	Numerus mul[tiplic]andus
22;58,42	9962239
4;46,36	
27;45,18	Productus numerus ε
Sinus	sinus declinationis
4656918	4639333
Declinatio igitur	australis
par. 27;38,28	

These values represent (see Figure 1.10a)

$$BC = 22;5\underline{8},42^\circ,$$

$$CD = 4;46,36^\circ,$$

$$BD = 27;4\underline{5},18^\circ,$$

$$\sin BD = 0.46569\underline{18},$$

$$\sin \angle CBA = 0.99622\underline{3}9,$$

$$\sin DE = \sin \angle CBA \cdot \sin BD = 0.46393\underline{33},$$

and

$$DE = 27;38,2\underline{8}^\circ.$$

Underlining indicates that changes were made in these numbers in the manuscript. It is not always possible to read the numbers written underneath, but in this case it is clear that Reinhold originally performed the entire computation under the assumption of minimum obliquity, 23;38°, and subsequently reworked his calculations under the assumption of a slightly larger obliquity, 23;29,1°.

Using 23;28° for the obliquity, the quantities listed above become

$$BC = 22;5\underline{7},42^o,$$

$$BD = 27;4\underline{4},18^o,$$

$$\sin BD = 0.46543\underline{43},$$

$$\sin \angle CBA = 0.99622\underline{6}9,$$

$$\sin DE = 0.46393\underline{47},$$

and

$$DE = 27;38,2\underline{9}^o.$$

Reinhold appears to have written these values for BC, BD, $\sin \angle CBA$, and DE.

The computation of the declination in Copernicus' second observation is similar. Reinhold first computes BC for an obliquity of 23;28° and then for 23;29,0°. The value 23;29,0° will not work for the first observation, as it produces 22;58,41° for BC. 23;29,1° will not produce the correct value for the second observation. According to Copernican theory, the obliquity of the ecliptic at the time of the observation should have been

$$\begin{aligned} \epsilon &= 23;40^o - 0;12^o \cdot \cos 166;18,47^o \\ &= 23;40^o - 0;11,40^o = 23;28,20^o. \end{aligned}$$

Between the first and second observations the obliquity should have been decreasing.

Reinhold omits one intermediate step in the determination of the declination. In order to solve ΔBDE, one needs $\angle CBA$. Assuming the obliquity is 23;29,1°,

$$\begin{aligned} \cos \angle CBA &= \cos AC \cdot \sin \angle CAB \\ &= \cos 77;24,43^o \cdot \sin 23;29,1^o \\ &= 0.0868463, \end{aligned}$$

and

$$\angle CBA = 85;1,4^o.$$

Then in right triangle BDE

$$\begin{aligned} \sin DE &= \sin \angle CBA \cdot \sin BD \\ &= \sin 85;1,4^o \cdot \sin 27;45,18^o \\ &= 0.99622\underline{17} \cdot 0.4656918 \\ &= 0.46393\underline{2}3. \end{aligned}$$

Reinhold has

$$\begin{aligned} \sin DE &= 0.99622\underline{39} \cdot 0.04656918 \\ &= 0.46393\underline{3}3. \end{aligned}$$

The value of DE, 27;38,28°, follows from both $\sin^{-1}$ $0.46393\underline{2}3$ and $\sin^{-1}$ $0.46393\underline{3}3$.

The Parallax from the Prutenic Tables

Using the values

$$z = 82;50^{o},$$

$$\bar{\eta} = 87;4,38^{o},$$

and

$$\gamma = 358;24,40^{o},$$

Reinhold's parallax table gives

$$\begin{aligned} P_1 &= c_4(2z) - c_7(2\bar{\eta}) \cdot c_3(2z) \\ &= 0;52,37^{o} - 0;59,50^{o} \cdot 0;2,13^{o} \\ &= 0;50,24^{o}, \end{aligned}$$

$$\begin{aligned} P_2 &= c_5(2z) - c_7(2\bar{\eta}) \cdot c_6(2z) \\ &= 0;62,33^{o} + 0;59,50^{o} + 0;3,26^{o} \\ &= 1;5,58^{o}, \end{aligned}$$

and

$$\begin{aligned} \pi &= P_1 + c_8(\gamma) \cdot (P_2 - P_1) \\ &= 0;50,24^{o} + 0;0,1^{o} \cdot 0;15,34^{o} = 0;50,24^{o}. \end{aligned}$$

Reinhold's value for the parallax is 0;52,2°.

THE SECOND PARALLAX OBSERVATION OF COPERNICUS
(Tables 1.5, 1.8)

Copernicus' second lunar parallax observation was made on the seventh day before the Ides of August (August 7), 1524, six hours after noon. The value of the observed lunar zenith distance has twice been changed, but this affects the final parallax, not the solar longitude and lunar position calculated here. The apparent time since epoch is 1524^{a} 234^{d} 18^{h}.

Neither of these values agree with those of Copernicus, who has written both 26;23° and 26;36° for the declination. It is not clear which number was written first, as the manuscript reads "part xxvi scrup xxxvi scrup xxiii," with the "scrup xxiii" crossed out (Ms.C.127r).

From Copernicus' table of declination,

$$\delta_m^* = (249;16^o) = 21;51,24^o$$

and

$$\delta_m^*(249;39^o) = 21;54,51^o.$$

Then

$$\delta_m^* + \beta = 21;51,24^o + 4;41^o = 26;32,24^o$$

in the first case, and

$$\delta_m^* + \beta = 21;54,51^o + 4;41^o = 26;35,51^o$$

in the second. Since the 23 is written xxiii, there is no mistaking it for 32, or xxxii. The 26;36°, however, agrees with one of Copernicus' values. Unlike the first observation, it appears that here Copernicus took the trouble to recompute the declination based on his corrected value of the lunar longitude.

The Parallax according to Ptolemy

Using the values

$$z = 80;55^o,$$

$$\gamma = 239;38^o,$$

and

$$\bar{\eta} = 97;5^o,$$

Ptolemy's parallax table gives

$$\begin{aligned} P_1 &= c_3(z) - c_7(\gamma/2) \cdot c_4(z) \\ &= 0;53,0^o + 0;44,10^o \cdot 0;10,12^o \\ &= 1;0,30^o, \end{aligned}$$

$$\begin{aligned} P_2 &= c_5(z) + c_8(\gamma/2) \cdot c_6(z) \\ &= 1;17,29^o + 0;43,34^o \cdot 0;24,15^o \\ &= 1;35,5^o, \end{aligned}$$

and

$$\begin{aligned} \pi &= P_1 - c_9(\bar{\eta}) \cdot (P_2 - P_1) \\ &= 1;0,30^o - 0;59,13^o \cdot 0;34,35^o \\ &= 1;34,38^o. \end{aligned}$$

Copernicus states that, according to Ptolemy, the parallax would have been 1;38°.

From Copernicus' table

$$\begin{aligned} P_1 &= c_3(2z) - c_7(2\bar{\eta}) \cdot c_4(2z) \\ &= 0;51,55^o - 0;59^o \cdot 0;2,5^o \\ &= 0;49,52^o, \end{aligned}$$

$$\begin{aligned} P_2 &= c_5(2z) - c_7(2\bar{\eta}) \cdot c_6(2z) \\ &= 0;61,46^o + 0;59^o \cdot 0;3,23^o \\ &= 0;65,6^o, \end{aligned}$$

and

$$\begin{aligned} \pi &= P_1 + c_8(\gamma) \cdot (P_2 - P_1) \\ &= 0;49,52^o + 0;44^o \cdot 0;15,16^o \\ &= 1;1,2^o. \end{aligned}$$

Copernicus has either $1;1\frac{1}{2}^o$, 1;5°, or 1° for the lunar parallax in this observation.

REINHOLD'S RECOMPUTATION OF COPERNICUS' SECOND OBSERVATION (Tables 1.9, 1.10)

In Julian years, the time since epoch of the second observation is 1523^j 219^d 18^h.

Solar Longitude and the Equation of Time

Reinhold finds, for the precession and anomaly of the equinoxes

	$\bar{P}$	θ
epoch	5;32,24	6;40,27
1500^j	20;55,54	2,37;21,31
23^j	0;19,15	2;24,45
219^d	0; 0,30	0; 3,46
	26;48, 3	2,46;30,29.

From the correction table

$$c_p(333;0,58^o) = 0;32,23^o.$$

The true precession is then

$$P = 26;48,3^o + 0;32,23^o = 27;20,26^o.$$

Reinhold finds the simple sun and solar anomaly

THE SECOND ECLIPSE
(Table 2.2)

The second eclipse reported by Ptolemy was observed in Babylon in the seventh year of Cambyses (Nabonassar 225), 17/18 Phamenoth. Ptolemy takes the time of mid-eclipse in Alexandria to be 1;50^h before midnight. The time since epoch is then 224^a 196^d 10;10^h.

Solar Longitude and the Equation of Time

The solar anomaly at the time of the observation was:

	$\bar{k}$
epoch	265;15
216^a	307;29, 7,16
8^a	358; 3,18, 2
180^d	177;24,51,39
16^d	15;46,12,35
10^h	0;24,38,27
0;10^h	0; 0,24,38
	44;23,32,37.

The equation of center is

$$c_s(44;23,33^o) = 1;36,23^o$$

from the table of anomaly, but 1;37,17^o accurately computed. The mean solar longitude is

$$\bar{\lambda}_s = 44;23,33^o + 65;30^o = 109;53,33^o,$$

and the true solar longitude is

$$\lambda_s = 109;53,33^o - 1;36,23^o = 108;17,10^o.$$

Ptolemy has 108;12° for the true solar longitude.

From the table of right ascension

$$\alpha(108;12^o) = 109;45,14^o.$$

If the mean solar longitude is adjusted by 0;5°, to correspond to Ptolemy's value of the true solar longitude, the equation of time is

$$\begin{aligned} E^o &= 109;45^o - 335;7^o - (109;49^o - 330;45^o) \\ &= -4;26^o, \end{aligned}$$

and

$$E^h = 0;4^{h/o} \cdot (-4;26^o) = -0;17,44^h.$$

Ptolemy uses –0;20^h. The corrected time since epoch is then 224^a 196^d 9;50^h.

Lunar Longitude

The mean lunar motions for the corrected time are

	$\bar{\lambda}_m$	$\bar{\gamma}$	$\bar{\varpi}$	$\bar{\eta}$
epoch	41;22	268;49	354;15	70;37
216^a	225;58,25,49	83;14,55,16	82; 1,57,51	278;29,18,33
8^a	315; 2, 9,50	349;44,59,49	109;42,17,41	316;58,51,47
180^d	211;44,55,40	191;41,48,53	221;16,59,26	34;20, 4, 0
16^d	210;49,19,36	209; 2,23, 0	211;40,10,37	195; 3, 7, 1
9^h	4;56,28, 6	4;53,57,43	4;57,39,37	4;34,17,30
$0;50^h$	0;27,27, 2	0;27,13, 7	0;27,33,40	0;25,23,50
	290;20,46, 3	27;54,17,48	264;21,38,52	180;28, 2,41

Ptolemy has 290;22° for the mean lunar longitude. Since he gives the true lunar longitude as 288;14°, his value of the lunar equation is

$$c_m = 290;22^\circ - 288;14^\circ = 2;8^\circ.$$

He gives 28;5° for the lunar anomaly, which presumably is the true anomaly. Since Ptolemy gives 108;12° for the true solar longitude, the mean solar longitude is approximately

$$\bar{\lambda}_s = 108;12^\circ + 1;36^\circ = 109;48^\circ.$$

If this is adjusted to the corrected time,

$$\bar{\lambda}_s = 109;48^\circ - 0;1^\circ = 109;47^\circ.$$

Then the elongation is

$$\bar{\eta} = 290;22^\circ - 109;47^\circ = 180;35^\circ,$$

not 180;28°. The lunar correction c_3 is then

$$c_3(1;10^\circ) = 0;10,18^\circ,$$

and the true lunar anomaly is

$$\begin{aligned}\gamma &= 27;54,18^\circ + 0;10,18^\circ = 28;4,36^\circ \\ &= 28;5^\circ.\end{aligned}$$

However, since Ptolemy's values of $\bar{\lambda}_m$ and $\bar{\varpi}$ do not agree with those found for the corrected time, there is no reason to expect his value of $\bar{\gamma}$ to agree either.

Ptolemy's value of the lunar equation is 2;8°, but his value of the true anomaly gives 2;11° for c_4 alone:

$$c_4(28;5^\circ) = 2;10,42^\circ,$$

$$c_5(28;5^\circ) = 1;5,32^\circ.$$

The value of the lunar equation is at least 2;13°:

$$c_m = 2;10,42^o + 1;5,32^o \cdot c_6(2 \cdot 180;28,3^o)$$
$$= 2;12,44^o$$

or

$$c_m = 2;10,42^o + 1;5,32^o \cdot c_6(2 \cdot 180;35^o) = 2;13,15^o.$$

The true lunar longitude is

$$\lambda_m = 290;22^o - 2;8^o = 288;14^o.$$

Ptolemy's value for the true argument of latitude is 262;12°. Thus his mean value is

$$\bar{w} = 262;12^o + 2;8^o = 264;20^o,$$

not 264;22°.

There is no consistent relation between Ptolemy's values for the mean lunar longitude and argument of latitude and the accurately computed values:

in the first eclipse

	Ptolemy	Accurately for 16;45^h
$\bar{\lambda}_m$	205;32	205;33
$\bar{w}$	79; 7	79; 4

and in the second eclipse

	Ptolemy	Accurately for 9;50^h
$\bar{\lambda}_m$	290;22	290;21
$\bar{w}$	264;20	264;22.

REINHOLD'S RECOMPUTATION OF THE SECOND ECLIPSE (Table 2.2)

Solar Longitude and the Equation of Time

Although Reinhold states clearly that 9;50^h is the corrected time since epoch, and the apparent time is 10;10^h, he uses 9;50^h to compute the solar anomaly:

	$\bar{k}$
epoch	265;15, 0
216^a	307;29, 7
8^a	358; 3,18
180^d	177;24,52
16^d	15;46,13
9^h	0;22,11
0;50^h	0; 2, 3
	44;22,44

Reinhold's value for the solar equation is 1;37°. Ptolemy's table of anomaly gives

$$c_s(44;22,44^o) = 1;36,22^o.$$

However, if the correct value for 42° is used, i.e., 1;33° rather than

$$c_s(44;22,44^o) = 1;36,58^o.$$

Accurately computed, the solar equation is 1;37,16°. The mean solar longitude is

$$\bar{\lambda}_s = 44;22,44^o + 65;30^o = 109;52,44^o,$$

and the true solar longitude is

$$\lambda_s = 109;52,44^o - 1;37^o = 108;15,44^o.$$

From Ptolemy's table of right ascension

$$\alpha(108;15,44^o) = 109;49,34^o.$$

Reinhold has 109;50°. The equation of time is then

$$\begin{aligned} E^o &= 109;50^o - 335;7^o - (109;52,44^o - 330;45^o) \\ &= 134;43^o - 139;7,44^o = -4;24,44^o, \end{aligned}$$

and

$$E^h = 0;4^{h/o} \cdot (-4;24,44^o) = -0;17,39^h.$$

Reinhold writes $0;17,30^h$ for the equation of time, but uses a corrected time of $9;50^h$, in agreement with Ptolemy.

Lunar Longitude

In computing the mean lunar motions, Reinhold writes separately the values for 9^h, $0;40^h$, and $0;10^h$. Since Ptolemy's table for hours only goes up to 24^h, the value for $0;50^h$ can be obtained by doubling the entry for 20^h and adding the entry for 10^h, adjusting the sexagesimal place.

	$\bar{\lambda}_m$	$\bar{\gamma}$	$\bar{\varpi}$
epoch	41;22, 0	268;49, 0	354;15, 0
216^a	225;58,26	83;14,55	82; 1,57
8^a	315; 2,10	349;45, 0	109;42,18
180^d	211;44,56	191;41,49	221;16,59
16^d	210;49,19	209; 2,23	211;40,10
9^h	4;56,28	4;53,58	4;57,40
$0;40^h$	0;21,58	0;21,46	0;22, 3
$0;10^h$	0; 5,29	0; 5,27	0; 5,31
	290;20,46	27;54,18	264;21,38

In accordance with Reinhold's practice of rounding, the values of $\bar{\lambda}_m$ for 16^d, and $\bar{w}$ for 216^a and 16^d should be increased by 0;0,1°.

Since Reinhold found the solar longitude for $9;50^h$, his elongation corresponds to the corrected time:

$$\bar{\eta} = 290;20,46^o - 109;52,44^o = 180;28,2^o.$$

From the lunar correction table

$$c_3(0;56,4^o) = 0;8,15^o.$$

Reinhold has 0;9°, producing

$$\gamma = 27;54,18^o + 0;9^o = 28;3,18^o.$$

Then

$$c_4(28;3,18^o) = 2;10,34^o,$$
$$c_5(28;3,18^o) = 1;5,28^o,$$
$$c_6(0;56,4^o) = 0;0,2^o,$$

and

$$c_m = 2;10,34^o - 0;0,2^o = 2;10,36^o.$$

Reinhold has 2;10,20°, less than the value of c_4 alone. The true lunar longitude is

$$\lambda_m = 290;20,46^o - 2;10,20^o = 288;10,26^o,$$

and the true argument of latitude is

$$w = 264;21,38^o - 2;10,20^o = 262;11,18^o.$$

LIST OF REFERENCES

Al-Battānī: cf. Nallino [1].

Anon. [1]: "Aurifaber", *The Encyclopaedia Britannica*, eleventh edition, v.2, New York 1910.

Archimedes [1]: *The Works of Archimedes*, ed. T.L. Heath, New York 1953.

Birkenmajer, A. [1]: "Le commentaire inédit d'Érasme Reinhold sur le *De Revolutionibus* de Nicolas Copernic", *Actes du Colloque international sur la science au seizième siècle*, Paris 1960, pp. 171–177.

Boyer, C. [1]: *A History of Mathematics*, New York 1968.

Brahe, Tycho [1]: *Astronomiae instauratae mechanica*, trans. H. Raeder, E. Strömgen, B. Strömgen, *Tycho Brahe's description of his instruments and scientific work*, Copenhagen 1946.

Britton, J.P. [1]: *On the Quality of the Solar and Lunar Observations and Parameters in Ptolemy's Almagest*, Yale Doctoral Dissertation [unpublished], 1967.

Burmeister, K.H. [1]: *Georg Joachim Rhetikus*, v.3, Wiesbaden 1968.

Copernicus, N. [1]: *De revolutionibus orbium caelestium.* Facsimile edition of Copernicus' manuscript, *Nikolaus Kopernikus Gesamtausgabe* I, Munich 1944.

Copernicus, N. [2]: *De revolutionibus orbium caelestium.* Critical edition, *Nikolaus Kopernikus Gesamtausgabe* II, ed. F. Zeller and C. Zeller, Munich 1949.

Copernicus, N. [3]: *De revolutionibus orbium caelestium.* Facsimile edition of the Nuremberg 1543 edition, *Copernicus de revolutionibus orbium caelestium libri vi*, Amsterdam 1943.

Delambre [1]: *Histoire de l'astronomie ancienne*, 2 v., Paris 1817.

Dreyer, J.L.E.[1]: *A History of Astronomy from Thales to Kepler (A History of the Planetary Systems from Thales to Kepler)*, 2nd ed. New York 1953.

Goldstein, B.R. [1]: "The Arabic version of Ptolemy's *Planetary Hypotheses,*" *Transactions of the American Philosophical Society*, new series, 57, 4, Philadelphia 1967.

Heath, T.L. [1]:*Aristarchus of Samos, the Ancient Copernicus*, Oxford 1966.

Hultsch, F. [1]: "Winkelmessungen durch die Hipparchische Dioptra," *Abhandlungen zur Geschichte der Mathematik* 9, Leipzig 1899.

Kepler, Johannes [1]: *Gesammelte Werke*, v.1, Munich 1938.

Kugler, F.X. [1]: *Sternkunde und Sterndienst in Babel*: 1, Munster 1907.

Manitius, K. [1]: "Die Parallaxen des Mondes und seine Entfernung von der Erde nach Ptolemäus," *Das Weltall*, 10. Jahrgang, Heft 3. 1909 November 1, Treptow-Berlin.

Nallino, C.A. ed. [1]: *Al-Battānī sive Albatenii, Opus Astronomicum*, 2 v., Publicazioni del Reale Osservatorio di Brera in Milano, No. 40, Rome 1903, 1907 (reprint, Frankfurt 1969).

Neugebauer, O. [1]: *The Exact Sciences in Antiquity*, 2nd ed., Providence 1957.

Neugebauer, O. [2]: "On the Planetary Theory of Copernicus," *Vistas in Astronomy* v. 10, Oxford etc 1968, pp. 89-103.

Neugebauer, P.V. [1]: *Tafeln zur astronomischen Chronologie II. Tafeln für Sonne, Planeten und Mond nebst Tafeln der Mondphasen für die Zeit 4000 vor Chr. bis 3000 nach Chr.*, Leipzig 1914.

Newcomb, S. [1]: "Refraction," *The Encyclopaedia Britannica*, eleventh edition, v.23, New York 1910.

Oppolzer, T. [1]: *Canon of Eclipses*, New York 1962.

Peurbach, G. [1]: *Theoricae novae planetarum*, ed. Erasmus Reinhold, Wittenberg 1553.

Proclus [1]: *Hypotyposis astronomicarum positionum*, ed. K. Manitius, Leipzig 1909.

Ptolemy [1]: Ptolemäus, *Handbuch der Astronomie*, deutsche Übersetzung und erläuternde Anmerkungen von K. Manitius, 2 v., Leipzig 1913 (reprinted with a foreword and corrections by O. Neugebauer, Leipzig 1962).

Ptolemy [2]: Claudii Ptolemaei, *Magnae constructionis, id est, perfectae coelestium motuum pertractationis lib. xiii.* Theonis Alexandrini, *in eosdem commentariorum lib. xi*, Basel 1538.

Ptolemy [3]: Claudii Ptolemaei, *Geographia* e codicibus recognovit, prolegomenis, annotatione, indicibus, tabulis instruxit, ed. Carolus Müllerus, v.1, part 2, Paris 1901.

Regiomontanus [1]: *Epitome in Almagestum Ptolemaei* [Venice 1496].

Reinhold, E. [1]: *Prutenicae Tabulae Coelestium Motuum*, Tübingen 1551.

Rome, A. ed [1]: *Pappus d'Alexandrie, Commentaire sur les livres 5 et 6 de l'Almageste; Commentaires de Pappus et Theon d'Alexandrie sur l'Almageste*, v.1, Rome 1931.

Rose, V. [1]: *Verzeichniss der Lateinischen Handschriften der Königlichen Bibliothek zu Berlin*, v.2, part 3, Berlin 1905.

Rosen, E. trans. [1]: *Three Copernican Treatises*, New York 1959.

Swerdlow, N. [1]: *Ptolemy's Theory of the Distances and Sizes of the Planets: A Study of the Scientific Foundations of Medieval Cosmology*, Yale Doctoral Dissertation [unpublished], 1968.

Swerdlow, N. [2]: "Hipparchus on the Distance of the Sun," *Centaurus* 1969, v.14, pp. 287-305.

Swerdlow, N. [3]: "Al-Battānī's Determination of the Solar Distance," *Centaurus* 1972, v.17, pp. 97-105.

Tuckerman, B. [1]: *Planetary, Lunar and Solar Positions A.D. 2 to A.D. 1649 at Five-day and Ten-day Intervals*, Philadelphia 1964.

Zinner, E. [1]: *Entstehung und Ausbreitung der Coppernicanischen Lehre*, Erlangen 1943.

Manuscripts

Reinhold, E. [2]: *Commentarius in opus Revolutionum Copernici*, Deutsche Staatsbibliothek, Berlin, Latin 2° 391.

INDEX OF NAMES

The names of Copernicus, Ptolemy and Reinhold are omitted in the index.

STUDIA COPERNICANA

I. Aleksander Birkenmajer, *Études d'histoire des sciences et de la philosophie du Moyen Âge*, Wrocław, 1970 (in French).

II. Mieczysław Markowski, *Burydanizm w Polsce w okresie przedkopernikańskim* [Buridanism in Poland during the Pre-Copernican Period] Wrocław, 1971 (in Polish with source materials in Latin).

III. Barbara Bieńkowska, *Kopernik i heliocentryzm w polskiej kulturze umysłowej do końca XVIII wieku* [Copernicus and Heliocentric Theory in Polish Intellectual Life till the End of the 18th c.] Wrocław, 1971 (in Polish).

IV. Aleksander Birkenmajer, *Etudes d'histoire des sciences en Pologne*, Wrocław, 1972 (in French).

V. *Colloquia Copernicana I. Études sur l'audience de la théorie héliocentrique (Toruń, 1973)*, Wrocław, 1972 (in English and German).

VI. *Colloquia Copernicana II. Études sur l'audience de la théorie héliocentrique (Toruń, 1973)*, Wrocław, 1973 (in English, German and Italian).

VII. Marian Biskup, *Regesta Copernicana*, Wrocław, 1973 (in Polish).

VIII. Marian Biskup, *Regesta Copernicana. Calendar of Copernicus' Papers*, Wrocław, 1973 (in Polish).

IX. Bronisław Biliński, *Najstarszy życiorys Mikolaja Kopernika z roku 1588 pióra Bernardina Baldiego.* [Nicholas Copernicus' Oldest Extant Biography Written in 1588 by Bernardino Baldi], Wrocław, 1973 (in Polish).

X. Karol Górski, *Lukasz Watzenrode. Życie i działalność polityczna (1447-1512)* [Lucas Watzenrode. Life and Political Activity], Wrocław, 1973 (in Polish).

XI. Grażyna Rosińska, *Instrumenty astronomiczne na Uniwersytecie Krakowskim w XV wieku* [Astronomical Instruments at the University of Cracow in the Fifteenth Century], Wrocław, 1974 (in Polish with source materials in Latin).

XII. Zofia Wardęska, *Teoria heliocentryczna w interpretacji teologów XVI wieku* [The Heliocentric Theory as Interpreted by Sixteenth-Century Theologians], Wrocław, 1975 (in Polish with source materials in Latin).

XIII. *Colloquia Copernicana III. Proceedings of the Joint Symposium of the IAU and the IAHS. Astronomy of Copernicus and Its Background (Toruń, 1973)*, Wrocław, 1975 (in English and French).

XIV. *Colloquia Copernicana III. Conférences des Symposia: L'audience de la théorie héliocentrique. Copernic et le développement des sciences exactes et sciences humaines (Toruń, 1973)*, Wrocław, 1975 (in English, French and German).

XV. *Witelonis Perspectivae Liber Primus. Book I of Witelo's* Perspectiva. An English Translation with Introduction and Commentary, and Latin Edition of the Mathematical Book of Witelo's *Perspectiva* by Sabetai Unguru, Wrocław, 1977 (in English with text edition in Latin).

XVI. *Science and History. Studies in Honor of Edward Rosen*, Wrocław, 1978 (in English, French and German).

XVII. *Nicholas Copernicus. Quincentenary Celebrations. Final Report*, edited by Zofia Wardęska, Wrocław, 1977 (in English).

XVIII. Jerzy Drewnowski, *Mikołaj Kopernik w świetle swej korespondencji* [Nicholas Copernicus in the Light of His Correspondence], Wrocław, 1978 (in Polish with source materials in Latin).

XIX. Jerzy Burchardt, *List Witelona do Ludwika we Lwówku Śląskim. Problematyka teoriopoznawcza, kosmologiczna i medyczna* [Witelo's Letter to Louis in Lwówek in Silesia, *De causa primaria poenitentiae in hominibus et de natura daemonum.* Problems of Epistemology, Cosmology and Medicine], Wrocław, 1979 (in Polish with text edition in Latin).

XX. *Georgii Joachimi Rhetici Narratio Prima.* Edition critique, traduction francaise et commentaire par Henri Hugonnard-Roche et Jean-Pierre Verdet, avec collaboration de Michel Lerner et Alain Segonds, Wrocław, 1982 (in French with text edition in Latin).

XXI. Erna Hilfstein, *Starowolski's Biographies of Copernicus*, Wrocław, 1980 (in English with Latin source materials).

XXII. Grażyna Rosińska, *Scientific Writings and Astronomical Tables in Cracow. A Census of Manuscript Sources, 14-16th Centuries*, Wrocław, 1984 (in English and Latin).

XXIII. *Witelonis Perspectivae Liber Quintus.* Book V of Witelo's *Perspectiva.* An English Translation with Introduction and Commentary, and Latin Edition of the First Catoptrical Book of Witelo's *Perspectiva* by A. Mark Smith, Wrocław, 1983 (in English with text edition in Latin).

XXIV. Grażyna Rosińska, *Optyka w XV wieku. Między nauką średniowieczną a nowożytną. Fifteenth-Century Optics. Between Medieval and Modern Science*, Wrocław, 1986 (in Polish with Latin texts and a comprehensive English Summary).

XXV. Pavel Spunar, *Repertorium auctorum Bohemorum provectum idearum post Universitatem Pragensem conditam illustrans,* vol. 1, Wrocław, 1985 (in Latin).

XXVI. Jane L. Jervis, *Cometary Theory in Fifteenth-Century Europe,* Wrocław, 1985 (in English with Latin texts).

XXVII. *Thomae de Wratislavia Practica medicinalis.* A Critical Edition of the *Practica medicinalis* of Thomas of Wrocław, Prémontré, Bishop of Sarepta (1297 - c. 1378) by Theodore Antry, O. Praem., Wrocław, 1989 (Latin text with English Introduction).

XXVIII. *Witelonis Perspectivae liber secundus et liber tertius.* Books II and III of Witelo's *Perspectiva.* A critical Latin Edition and English translation with Introduction, Notes and Commentaries by Sabetai Unguru, Wrocław, 1991 (in English with text edition in Latin).

XXIX. *Witelona Perspektywy księga II i księga III.* Przekład na język polski ze wstępem, opracowaniem i komentarzami. Redaktor przekładu: Witold Wróblewski. Przekład, opracowanie i komentarze: Lech Bieganowski, Andrzej Bielski, Roman S. Dygdała, Witold Wróblewski, Wrocław, 1991 (in Polish with English summary).